HOW TO OBTAIN
ABUNDANT CLEAN ENERGY

HOW TO OBTAIN
ABUNDANT
CLEAN ENERGY

Linda Baine McGown

California State University
Long Beach, California

AND

John O'M. Bockris

Texas A & M
College Station, Texas

PLENUM PRESS · NEW YORK AND LONDON

Library of Congress Cataloging in Publication Data

McGown, Linda Baine.
   How to obtain abundant clean energy.

   Includes index.
   1. Power resources. I. Bockris, John O'M., joint author. II. Title.
TJ163.2.M28   621   79-24468
ISBN 0-306-40399-4

Preface

The Three Mile Island accident, the periodic occurrence of long gas lines until sufficient price increases are achieved, our dependence on foreign powers for a large fraction of our energy supply, and continual controversy in our government and concerned public groups all point to the real presence of an energy crisis. Even the government has finally publicly acknowledged the fact that our present fuel sources will run out soon enough to be of concern to us now.

This knowledge should raise many questions in the minds of our citizens. When will our sources run out, or at least become too expensive to afford? What could replace them, and which alternatives are best?

When we hear about these matters in the news media, we get many contradictory opinions, mainly concerning coal and nuclear energy. Most of us realize that the reintroduction of coal on a massive scale would give rise to considerable pollution difficulties. Many people are also waking to an awareness of the dangers of nuclear reactors.

When we turn to scientists for answers, we find that each one seems to advise us that a single path to new energy sources is the best—of course, each scientist tends to advise his own path.

In this book, we have attempted to objectively describe the advantages, disadvantages, and possible dangers of each energy alternative. It is likely that there will be a multipath development of energy resources in the future. Each of the considered sources has some part to play, and for the next hundred years we shall exist upon a mixture, increasingly made up of *renewable energy resources*, i.e., those which will not run out.

We stress in our book not only the technological aspects, but also considerations of *time* and *money*. These latter factors must be dealt with here because technological feasibility becomes a realistic possibility only if the idea can be transformed into an affordable commodity in time to replace our exhausting fuels.

There is a corollary to the consideration of time and money: politics. We have broken through the traditional nonpolitical stance of most scientists in our decision to discuss this aspect, too, and to shine some light upon the inertial weight of the vast investments of the oil companies and those of the rich and hence powerful OPEC nations in the old energy technology.

One of us (JOMB) would like to express thanks to his wife, Lily

v

Bockris, for discussions and collaboration in the early stages of writing this book.

California State University,   Linda Baine McGown
Long Beach, California

Texas A&M University   John O'M. Bockris
College Station, Texas

Contents

PART I. THE APPROACHING ENERGY DISASTER

Chapter 1. What Is Energy? 3
  Energy Is That Which Makes Things Go 3
  Classification of Energy 3
  Potential Energy 4
  Turning Potential Energy to Useful Energy 6
  Kinetic Energy 6
  Potential Energy in Fuels 7
  How the Energy Got Inside the Fossil Fuels 9
  A Nasty Realization 9
  The Conversion of the Various Forms of Energy to Forms
      Which Can Be Used by People 10
  The Vital Importance of Having a Good Transducer (Converter) 10
  Efficiency of Energy Conversion 11
  The Predominance of Cost 11
  The Importance of Compromise 12
  The Importance of the Time Left 12

Chapter 2. What Part Does Energy Play in Our Lives? 13
  Introduction 13
  Energy Determines the Kind of Lives We Lead 13
  The Standard of Living 14
  Income, The Standard of Living, and Energy Per Person 15
  Energy, Comfort, and Efficiency 15
  The Divisions of Energy Use 15
    Household Energy 17
    Transportation Energy 17
    Industrial and Commercial Energy 19
  One (Avoidable) Negative Aspect of a High-Energy Life: Pollution ... 20
  Another Negative Aspect of the Present Energy System: The Coming Exhaustion
    of Our Present Sources 22

Chapter 3. How Do We Get Our Energy Right Now? 25
  Introduction 25
  A Summary of Our Present Energy Sources 25
  Making Electricity 26
  Fossil Fuels 29
  How We Get Our Energy in Industry 34
  Problems of the Near Energy Future 36
  How Energy Is Transported Now 37
  Our Finite Energy Sources 38

Chapter 4. Shall We Run Out of Energy in Our Time? 39

Present Fuels Are Limited 39
How Do We Estimate the Fossil Fuels Left? 39
   The French Farmer 40
Resources Exhaust Suddenly 41
The Date of the Maximum Production Rate—The Effective End 42
When Will The Maximum Production Rate Be Reached? 43
The Effect on the Exhaustion Date of Rising Living Standards and Population Growth ... 45
How the Rate of Production of a Resource Will Vary with Time 45
The Critical Year of the Maximum 46
The Effect of Change in Price of a Resource on Its Use Rate 47
How an Estimate of 1500 Years of Coal Can Be Revised to Less than 30 Years 48
When Shall We Run Out of Oil and Natural Gas? 49
Running Out of Coal, Too 50
More Pollution? 51
What Would Happen If We Stopped the Growth of Our Economies? 51
Is Conservation the Answer? 52
Shall We Run Out of Fossil Fuels in Our Time? 53

Chapter 5. How Long Does It Take to Develop and Build Up a New Technology? 55

Introduction 55
Stages in Developing a New Technology 56
   The First Stage—Dreams 56
   The Second Stage—Fundamental Research 57
   The Third Stage—Developmental Research 57
   The Fourth Stage—Commercialization 57
The Cost of the Various Stages of Developing Technology 58
Historical Guidelines Tell Us How Long Technologies Take to Develop 59
It is Likely to Take 25–50 Years (1–2 Generations) to Develop New Energy Sources 59
Monetary Aspects of Building a New System 60
A Near Thing? 60

PART II. ALTERNATIVES: WHAT COULD REPLACE OUR EXHAUSTING FUELS?

Chapter 6. The Dream of Cheap, Clean Atomic Energy 69

The Beginning of the Dream 69
The Dream Continues 70
A Quantitative Comparison 71
The Dream Fades? 71
Types of Atomic Reactors 72
The Fission Reactor 72
   Uranium—The Active and the Stable Forms 73
   How Fission Reactors Work 73
   The Difficulty with Nonbreeder Fission Reactors: Exhaustion 75
Will the Breeder Reactor Save the Situation? 77
The Drawbacks of Breeder Reactors 79

Chapter 7. Fission Reactors—What Can Go Wrong 81

Introduction 81
Fission Reactor Accidents—A Few Close Calls 81
   What Can Go Wrong? 81
   The Brown Ferry Mishap—Human Error in Action 83
   Three Mile Island—"But No One Was Killed" 84
Biological Hazards of Nuclear Radiation—"Where Does It Hurt?" 84
   What Increases in Background Radiation Are Expected if
      We Switch to Atomic Power? 85
Cancer-Causing Cell Damage by Radioactive Substances 85
   The Dose Rate Affects the Extent of Biological Damage 87
Dumping Radioactive Garbage in Whose Backyard? 89

Chapter 8. Dreaming About the Future:
Abundant Clean Energy from Atomic Fusion 91

Fusion? 91
Fusion: Another Atomic Alternative 91
The Advantages of Fusion as an Energy Source 92
The Diluteness of Deuterium 92
Is the Fusion Concept Utopian? 92
Atomic Fission and Atomic Fusion 93
But Shall We Actually Be Able to Attain the Fusion Process? 94
Difficulties in Realizing a Controlled Atomic Fusion Process 95
Plasmas 96
The Difficulty of Containing a Plasma in a Bottle Is That It Escapes 96
Another Possible Method of Getting Energy from Fusion 99
What Has Been the Progress of the Laser Method for Fusion? 101
Time 104
Is Fusion the Best Energy-Producing Prospect of Them All? 104
Fusion Compared with Other Abundant Clean Energy Sources 105

Chapter 9. The Most Available Energy Source: The Sun 107

Energy from the Sun 107
The Sun's Expected Life 107
How Much of the Solar Energy Radiated from the Sun Reaches the Earth? 109
Solar Energy Reaching the Earth 110
Solar Energy per Person 111
How Much Energy Does the Average Person Consume? 112
Trying to Allow for the Future: How Much Energy Will Be Needed
   by the Years 2000 and 2050? 113
Solar Energy is Dilute 114
How Much of the Earth's Surface Can Be Used for Solar Collectors? 114
What Will Be the Efficiency of Collection of Solar Energy? 115
Will the Amount of Solar Energy Which We Could Collect Be Enough to
   Supply Our Total Energy Needs? 116
Shall We Need More Energy than We Have Calculated Above in the Further Future? 117
To What Medium Shall We Convert Solar Energy for Use? 117
How Will We Get Solar Energy from the Places Where It Is Easily Available
   to Where It Is Needed? 119

How Much Time Will Be Needed to Make the Solar Collectors? 120
If Solar Energy Is Readily Available, Why Wasn't Its Collection
    Developed Many Years Ago? 123
The Strange Situation of Countries with Solar Energy That Do Not Collect
    the Solar Energy 123

Chapter 10. Converting Solar Energy to Useful Fuel 125
Introduction 125
The Solar Spectrum 125
The Photovoltaic Method of Converting the Sun's Energy to Usable Energy on Earth 127
    Is the Photovoltaic Method of Collecting Solar Energy Too Expensive? 129
    Thin Film Photovoltaics 131
    Photovoltaic Collectors in Orbit 132
The Mirror Concentrator Method 134
Ocean Thermal Energy Collectors (OTEC) 135
To What Extent Have the Methods Described in this Chapter
    Actually Been Built and Used? 137

Chapter 11. Household Energy from the Sun 141
Introduction 141
The Production of Hot Water 143
Space Heating of Houses 143
Space Cooling 146
Household Electricity 148
What Do We Do When the Sun Goes Down? 149
Will Solar Energy for Households Be Commercially Available Before Oil Runs Out? 151

Chapter 12. Transport and Industry Run On Electricity and Hydrogen ... 153
Introduction 153
Running Cars in the Post-Fossil-Fuel World 154
Could We Run Cars on Batteries, Charged by Electicity? 156
    Cars Run on Batteries Which Work with Lead Electrodes 156
    The Sodium-Sulfur Battery 157
A Source of Energy to Charge Batteries for Electric Cars 158
Hydrogen-Driven Cars 159
Hydrogen-Driven Planes 160
Fuel Cells: How to Get Back Electricity from Hydrogen Derived from the
    Energy of Solar Radiation 161
The Poor Efficiency of Ordinary Engines 163
The Better Efficiency of Electrochemical Engines 164
Running Industry on Hydrogen 164
Foods from Hydrogen 165
Metallurgy 167

Chapter 13. Tides, Geothermal Heat, and the Big Winds 169
The Big Winds: How They Could Be Used to Give Hydrogen Fuel and
    Electrical Energy for Cities 169
Could Wind Be a Reliable Source of Energy on a Large Scale? 169
Could Wind Generators Produce Household Electricity? 170

More Wind Energy Estimates 171
The Big Winds 171
Winds at Sea Are Stronger 174
How Would Wind Energy Be Stored on a Massive Scale? 174
Would Massive Wind Power Be a Practical Proposition? 177
It Is Always the Cost that Counts 181
Energy from the Tides? 181
Energy Beneath Our Feet 185
How Hot Rock Geothermal Energy Might Become Practical 187
Difficulties in the Attainment of Hot Rock Geothermal Energy 187
Low-Grade Geothermal Energy 190
Summary of the Prospects of Geothermal Energy 190

Chapter 14. Energy Storage and Transmission 193
Energy Carriers: A Choice Among Three 193
Sources and Media 194
What Are the Possible Media (Carriers of Energy)? 194
Pros and Cons of the Various Media 195
Transmitting Energy Over Long Distances 196
Why Long Distance Electric Transmission Is Not Acceptable 197
Hydrogen Could Help Reduce the Cost of Sending Energy Over Long Distances 198
Very Long-Distance Transmission of Energy 198

PART III. THE HYDROGEN ECONOMY

Chapter 15. Methods of Mass-Producing Hydrogen 205
Introduction 205
The Cyclical Chemical Method for Producing Hydrogen 205
Disadvantages of the Cyclical Thermal Method 207
The Electrochemical Method of Obtaining Hydrogen 208
Getting the Energy Back from Gaseous Hydrogen at the User Terminal 211
What of Homes and the Electricity We Now Use in Them? 212
The Advantages of Using Fuel Cells 213

Chapter 16. The Storage of Abundant Clean Energy 215
Introduction 215
Methods of Storing Energy 216
Storage of Energy in the Form of Heat 217
Storing Energy in Its Electrical Form 218
The Pros and Cons of Heat and Electrochemical Storage 218
Storing Gaseous Energy Underground 219
What Methods Will Be Most Used in Our Time for Energy Storage? 219

Chapter 17. Beyond the Hydrogen Economy: Some Futuristic Ideas 223
Concepts for the Next Few Hundred Years 223
Concepts of the Next Few Thousand Years 223

PART IV. EXTRASCIENTIFIC CONSIDERATIONS

Chapter 18. The Politics of Survival 233
    Introduction 233
    The Direction of Major Research Funding Depends upon Politicians 234
    What Does Survival Mean? 235
    The Idea of Vested Capital 236
    The Tobacconist 236
    The Politician's Dilemma 237
    Economies Cannot Expand Forever—When Will Growth Stop? 238
    Energy Disaster?—People, Politics, Government Funds, and Research 239

Chapter 19. Answers 241

Glossary 243
Index 253

HOW TO OBTAIN
ABUNDANT
CLEAN ENERGY

Part I

The Approaching Energy Disaster

This part of the book begins an attempt by two scientists to answer, to explain, or in some cases to just discuss the following questions, which are of concern to each of the citizens of all the industrialized nations of the world:

What energy crisis?
What energy disaster?
Is oil really running out?
What *is* energy?
Can pollution be avoided?
What about atomic energy?
Is fission safe?
Are breeder reactors the answer?
Fusion?
Is solar energy on a massive scale feasible and practical?
What about coal?
How long will it take to develop alternatives, and how much will they
   cost?
How much time is left before the alternatives must be in full use?
What effect will the energy crisis have on our present lives?
What effect will the impending energy disaster have on the lives of our
   children?

1

What Is Energy?

Energy Is That Which Makes Things Go

Energy is behind and within all things that move or that are capable of motion. It alone causes work and change to take place. It is essential for biological existence, for the atmosphere's warmth which embraces us, for the light by which we see, and for the power by which we move. In fact, even matter itself can be viewed as a form of stored energy.

So, before beginning a discussion of the present energy crisis of modern society and the impending energy disaster which threatens us, it is essential to realize that energy is more than coal, or gasoline, or "get up and go" in a person. It is involved in every aspect of our lives. The story of the energy crisis is one of the difficulty in obtaining energy in a form which is useful to industrialized society. Since manpower is being replaced more and more by machine power, we have made a transition from reliance on food energy (which moves the worker) to electricity or steam, which moves the machines. Hence, when a modern person thinks of energy, he is usually referring to gas for his car, electricity for his home, or fuel for his factories. Energy manifests itself in many other forms, and in this book we shall explore ideas for making some of these other forms useful to industrialized people.

Classification of Energy

As stated above, energy manifests itself in many different forms. Among the different types of energy, familiar to most people, are the following:

Heat: A reflection of how fast atoms and molecules are moving in a substance, measured in terms of temperature [i.e., degrees Celsius (C) or Fahrenheit (F)] or calories.

3

Electricity: Produced by a flow of electrons from one point to another, i.e., an electric current. The force making the electrons flow is measured in terms of volts, the current in amperes (amps), and the power in watts.

Gravitational: The attraction between two masses, such as the pull of the earth on a rock or a person, or the pull of the moon on the oceans (causing tides).

Radiation: The most familiar example of radiative energy is light (e.g., sunlight, incandescent bulbs, fluorescent lamps, candle light). Other common forms include x-rays (used for medical diagnosis and treatment, and *very* dangerous), microwaves (used in microwave ovens for cooking), radio waves (used for communication), and gamma rays, which are produced by nuclear reactions in the sun and, recently, on earth.

Magnetic: Such as that produced by the earth's magnetic poles.

There is, in addition, a more general way of classifying energy, according to whether it is being used or being stored. *Kinetic energy* is the energy of motion, in other words, work is being done (a mass is being moved through a distance in space). *Potential energy* is stored energy, as in fuels or batteries. The ability or potential to do work exists, but is not currently being exploited. Thus, in these terms, the energy crisis of modern society involves a search for new sources of potential energy, which can be converted to kinetic energy to do useful work for mankind, to replace those which are being depleted.

POTENTIAL ENERGY

In its ordinary use, the word "potential" means something which is in reserve. Thus, it is something which exists, but is not yet in use. Consider a spring which is wound up. The application of such a spring in a watch causes movement. The taut nature of the coil in the spring evidently harbors energy, something which can manifest itself in the movement of the gears of the watch. The energy in the spring can be called potential, or stored, energy. It becomes movement energy (or kinetic energy) when necessary.

Another example is an object on top of a cupboard. This object also has potential energy. If we push the object over the edge of the cupboard, it falls to the ground. It must have had some stored energy before it started to travel the distance from the cupboard top to the ground level. Evidently, by virtue of its location, the object had energy which showed up when it fell, i.e., it moved. We say that the object had stored, or potential, energy, caused by the earth's gravitational field (Table 1.1).

A more abstract concept is the potential energy contained inside *mole-*

TABLE 1.1. POTENTIAL ENERGY OF A MASS OF 1 KILOGRAM
AT VARIOUS HEIGHTS ABOVE THE EARTH'S SURFACE

Height (meters)	Potential energy (joules)
1	9.81
4	3.9×10^1
10	9.8×10^1
4×10^1	3.9×10^2
1×10^2	9.8×10^2
1×10^3	9.8×10^3
1×10^5	9.8×10^5

cules. From the point of view of the general subject in this book, finding out about energy, the energy inside molecules is important because it is the source of energy we use at present.

Molecules consist of atoms bound together through links, or bonds, and these bonds can be regarded as springs, vibrating to and fro. Thus, if we consider the energy of a bond, represented for the moment by two atoms vibrating back and forth like two spheres at the end of a dumbbell, a little thought shows that the total energy of the bond must be partly potential, and partly kinetic, and the proportions of each vary with the position of the atoms about the bond (Figure 1.1). When the atoms are not moving, i.e., when they are either as far apart as possible, and ready to move back toward each other, or when they are as close as possible, and ready to turn and move away from each other, all of the energy is potential (there is no motion). When the atoms are moving fastest, either toward or away from each other, there is only kinetic energy. At any other point along their vibrational path, there is some combination of potential and kinetic energy, depending on the position of the point.

Thus, each chemical bond must contain some potential energy (Table 1.2), although the amount varies with position. The total energy is constant, and the ratio of kinetic and potential energies varies as the atoms oscillate back and forth.

FIGURE 1.1. We get energy from fuels like hydrogen because they produce explosive reactions with oxygen. Water contains less energy than the original hydrogen and oxygen. The excess energy is released as heat, and it is this kind of heat which now provides the world's energy.

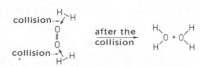

TABLE 1.2. THE AVERAGE POTENTIAL ENERGY CONTAINED IN VARIOUS CHEMICAL BONDS

Bond	Potential energy (kJ/molecular unit)
Hydrogen–hydrogen	4.4×10^2
Oxygen–hydrogen	4.3×10^2
Carbon–carbon	6.0×10^2
Nitrogen–oxygen	6.3×10^2
Potassium–iodine	3.2×10^2
Copper–silver	1.7×10^2
Iodine–iodine	1.5×10^2
Carbon–nitrogen	7.7×10^2

TURNING POTENTIAL ENERGY TO USEFUL ENERGY

In what way can we use stored chemical energy to perform work, for example, force a car forward? Consider the burning of hydrogen with oxygen to form water. Together, the hydrogen molecule and the oxygen molecule have a certain total potential energy stored in their bonds. Water, the product of the reaction, is made out of the hydrogen and oxygen molecules, and consists of two hydrogen atoms each bound to an oxygen atom. Each hydrogen–oxygen bond has a potential energy, and the sum of these energies is less than the sum of the potential energies of the original hydrogen and oxygen molecules. So, during the reaction, energy is *released* as water is formed (Figure 1.1).

Energy itself is indestructible, or conserved, only changing in form and not quantity. The energy stored in the bonds of the hydrogen and oxygen molecules is not lost. It simply changes form, and is released as heat energy. The potential energy has been transformed into kinetic energy—the water molecules move faster—which is manifest as heat. Chemical potential energy has been transformed into kinetic heat energy, which can be used to do work (Table 1.3).

KINETIC ENERGY

Kinetic energy is the easiest kind of energy to understand. We can *feel* it (and hear it), for example, in the collision of two masses in motion. One car moving along a street at a certain velocity collides with another car, and in a couple of seconds both cars are stationary. But metal has been bent, i.e., the kinetic energy of the moving vehicles has been converted by the collision into potential energy, energy which caused the stretching of the bonds between the atoms of the metal in the car body, and caused the metal to stretch and turn, and to split and break. Here, the relationship between the literal energy or go-power, and the idea of potential energy to which it is convertible in the collision is again clear. The kinetic energy of the moving cars causes the potential energy between the bonds of the atoms, holding them together in chemical bonds, to be overcome, breaking apart the metal of the car (Figure 1.2).

POTENTIAL ENERGY IN FUELS

Energy is stored inside the molecules which make up fuels. It is the *potential* within the chemical bonds of the fuel molecules. When we unite these molecules with oxygen, during combustion, the potential energy in the molecules produced is less than in the original molecules (just like in the example of water made from hydrogen and oxygen discussed above), and again the difference in energy is released as heat, which we can use to do work.

For example, when you press your foot against the accelerator of a car, more gasoline is mixed with molecules of oxygen and burned to form more products (automobile exhaust). More heat is also produced, and it is this heat which causes the expansion of the mixture to press against the piston, which then drives the transmission to rotate the gears and drive the wheels. Thus, the car moves forward. Potential energy has become energy of motion (Figure 1.3).

TABLE 1.3. THE KINETIC ENERGY OF VARIOUS SITUATIONS

Situation	Rough value of typical mass (kg)	Typical velocity (m/sec)	Approximate energy (J)
A hydrogen molecule in a gas at normal pressure	3×10^{-27}	10^3	2×10^{-21}
A 1-kg weight after falling 3 m	1	7×10^{-1}	3×10^{-1}
A bird in flight	0.1	3×10^1	5×10^1
A ball in flight	0.2	3×10^1	1×10^2
A man walking	7.10^1	2	1×10^2
A car at speed	3.10^3	3×10^1	1×10^6
A Boeing 747 in flight	3.10^5	3×10^2	1×10^{10}
A space vehicle in orbit	3.10^4	8×10^3	1×10^{12}

FIGURE 1.2. Photograph of a totaled car: *Kinetic* energy being converted to *potential* energy.

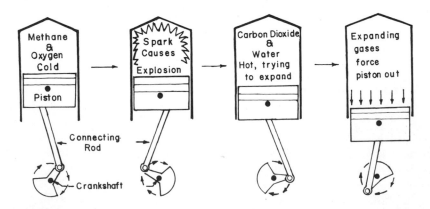

FIGURE 1.3. Conversion of chemical potential energy to kinetic energy. The connecting rod feels force and moves. (Force × distance = energy spent.)

How the Energy got Inside the Fossil Fuels

"In the beginning was the Word," says the Bible, but a more pertinent statement would be, "In the beginning there was much sunlight." (Later, in Chapter 9, this statement will be modified further.) The light energy was stored in the fossil fuels, and the mechanism by which this happened offers a good illustration that all energy is interconvertible. It has been generally accepted that the fuels we now use, which are found deep in the earth, were at one time, many millions of years ago, formed by a process called photosynthesis. This process occurs in most plants (and in *no* animals) and uses two of our most abundant molecules, carbon dioxide and water vapor (both found in the air) to form cellulose, which is the basic structure from which our plants and trees are made. The energy for the reaction between carbon dioxide and water to form cellulose *is provided by light*, which is supplied by the sun.

So, wood has solar energy stored within its chemical bonds in the form of potential energy. When we burn wood to form carbon dioxide and water, the stored energy is converted into useful heat. This heat can be used to convert water into steam, which will drive generators and produce electricity. But it is, in effect, stored solar energy from the sunlight of an earlier epoch.

All of the wood that formed the trees millions of years ago, which grew under light, died, and decayed, gradually underwent several complicated chemical processes, some of which finally produced coal, petroleum, or crude oil.

Thus, fuel molecules such as these came originally from cellulose, and this comes from carbon dioxide and water and solar energy. All of the energy we are using in our petroleum fuels today is *stored solar energy* given us by the photosynthesis reactions of millions of years ago, and the energy which *causes* this reaction is the energy of sunlight.

A Nasty Realization

This leads to an interesting and essential point about our present age. Until the middle of the last century, we had hardly used any of our fossil fuel resources to get energy. Then, in the 19th century, came the idea of using oil as a fuel. Oil is easier to deal with than coal (oil flows down pipes and does not have to be transported in trains), and quickly became heavily relied upon as an energy source. Later, natural gas (methane) was discovered, and it was found that this was still easier to use. Since gases flow more easily than liquids, it is simpler to transport natural gas a long distance through pipes than oil. Rapidly, we developed our technological civilization, based upon the potential energy stored in the molecules of these fossil fuels.

What we did not realize soon enough is that the "bank deposit" of our fossil fuels has a finite number of energy units in it, i.e., that the *total amount of fossil fuels* which we had been burning up with abandon, running our civilization at full blast, *is limited*. The pools of oil will run dry, and eventually even the big seams of coal which are underground will get used up.

These thoughts are the basis for the title of this book. We have to transfer our dependency on *exhaustible* energy sources (the fossil fuels) to *inexhaustible* or *renewable* ones. A chief possibility is direct solar energy, which can provide us with heat and electricity, if we can engineer conversion systems at sufficiently low cost.

The Conversion of the Various Forms of Energy to Forms Which Can Be Used by People

Since all energy, despite all of its different forms, is the same thing, merely *appearing* different, like an actor dressed in many costumes, we should be able to convert it from one form to another. Thus, if we have some heat, we ought to be able to change it into electricity, and if we have potential energy, we ought to be able to make it become light energy, and so on.

All of these changes are realizable. In everyday existence, when we switch on the electric light, the current in the wires comes from the generator at the local electric plant. This generator is being made to turn by a piston moving in a cylinder, with the piston pushed to and fro by gases expanding from the heat released by the combination of molecules (burning of oil or coal—see Figure 1.4), to give energy in a way we have seen above. Originally, all of the energy we are discussing came from the radiation of the sun.

The Vital Importance of Having a Good Transducer (Converter)

It should be clear from what we have said that good transducers, i.e., good energy converters, are vital to the future of industrialized civilization. *Raw* forms of energy (heat, light, etc.) have limited use because of a lack of flexibility. We need to have energy in an accessible form. It has to be processed, i.e., put into a form which we can send from place to place, turn on and off, power-regulate, etc. One such convenient form of energy, we might call it an energy *medium*, is electricity, and another is the refined form of oil, gasoline.

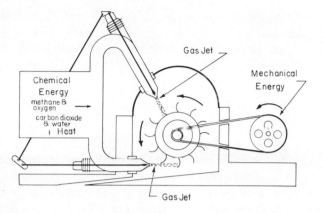

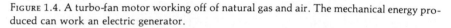

FIGURE 1.4. A turbo-fan motor working off of natural gas and air. The mechanical energy produced can work an electric generator.

EFFICIENCY OF ENERGY CONVERSION

Conversion of energy from one form to another always involves some loss of energy during the process. This loss is not only due to the energy involved in the manufacture and maintenance of the transducer, but also to the conversion process itself. By energy loss, we do not mean energy is destroyed, but simply that it is given off in a way in which it can't be collected for useful work. There are no transducers which will take a number of units of energy in form 1 and give out the same number of energy units in form 2. All transducers *lose* energy in the process of conversion; i.e., they demand a price above that of their cost and maintenance. Often, this energy is lost as heat. If it were only a matter of a few per cent of the total, it might not worry us, but it often comes to 60–70%, or even more, of the energy we put into the transducer initially. Hence, one of the things we have to know about each transducer is its *efficiency*. If the output of energy in form 2 is only 1% of the energy input in form 1, the transducer is not very attractive.

THE PREDOMINANCE OF COST

We have discussed in this chapter what energy is, stressing that energy is the entity behind all movement and that it occurs in many forms. Also, we have mentioned the interconversion of one form of energy to another, and how this always entails some energy loss along the way.

At this point, financial considerations must be introduced. The only useful energy source is one which is affordable. Hence, along with the scientific and technological aspects of energy is the economic aspect. How much would it cost to maintain a certain expected standard of living using a particular energy source and transducer? In fact, it is ultimately the *price of energy that governs the material standard of life.*

The Importance of Compromise

The choice of our energy source depends on the amount and availability of our supply (coal, atomic, solar). The choice of a transducer or conversion scheme (e.g., solar radiation → heat → steam → generator → wires → motor, or coal → steam → heat engine → mechanical energy) depends on a number of factors, including the degree of pollution associated with its use, the efficiency of the transducer, and the type of energy which will be produced. Does it need storage? If so, is the form of energy difficult to store, like mechanical energy, or easy to store, like oil?

The best compromise must be made, taking all of these factors into account. This is obvious. However, in order to come up with practical solutions when talking about good ideas and marvelous new concepts, *cost* must be evaluated at each step.

Cost evaluations will not be very accurate upon first examination of a new idea, of course. Finally, however, the public will want the *cheapest* method, and an engineer's efforts will be fruitless unless he can provide a *cost-acceptable* way of producing the energy needed to provide the public with their desired lifestyle.

The Importance of the Time Left

One last consideration for this chapter is time. It takes *time* (not months or years, but several decades at least) to research, develop, and commercialize new concepts. There are two major questions to be asked. How long will our present fuels last? How long will it take to research and develop another source of energy (a source both abundant and clean), in a cost-acceptable fashion, so we can *use* it?

We had better make the latter time shorter than the former.

2

What Part Does Energy Play in Our Lives?

INTRODUCTION

Even a person in a primitive community, living without machines, depends on energy in his life, for energy is connected with all of his movements, his eating, his breathing. For others, living in modern, industrialized communities, energy is used in factories, transport, lighting, heating, cooling, communication, and recreation. Gasoline-powered automobiles are used for transportation. Rooms are heated with energy produced from oil and lighted with electricity produced from coal. Kitchens are equipped with products from factories made by machines driven by natural gas, oil, or electricity, and the appliances consume these fuels themselves.

Not only does energy get used in these direct ways, for transportation and household uses, but, in addition, there is a certain amount of energy which goes into the making of every manufactured product. A kilogram of margarine contains an amount of energy, not only in the sense that it can be converted in the body to energy which activates our movements, but also in its own production, for this occurs by machines which use energy. Table 2.1 illustrates this point.

ENERGY DETERMINES THE KIND OF LIVES WE LEAD

All of our activities are connected with energy use. When we walk across a room, we use energy from biochemical energy production reactions of the food we eat and the air we breathe. When we read, we are using the energy from the light bulb, which came from the electricity passing through its wires, which in turn was produced from potential energy stored in an

TABLE 2.1. TYPICAL ENERGY CONTENTS OF MATERIALS
AND MANUFACTURED PRODUCTS

Material	Energy[a] (MJ/kg)	Percent of cost of product attributable to energy
Metals		
Steel (various forms)	25–50	30
Aluminum (various forms)	60–270	40
Copper	25–30	5
Magnesium	80–100	10
Other products		
Glass (bottles)	30–50	30
Plastic	10	4
Paper	25	30
Inorganic chemicals (average value)	12	20
Cement	9	50
Lumber	4	10

[a]These are typical values of energies consumed in the manufacturing process. The actual value depends on the purity, form, manufacturing process, and other variables.

electrochemical system. When we drive to work, we are using the solar energy stored millions of years ago by photosynthesis in plants, which later would become oil. When we use a tea kettle or a coffee pot, it consists of steel made by a high energy process. Wars are fought with planes, tanks, and guns, using chemical and electrical types of energy. And news of the war is brought to us on electrically run televisions and radios.

THE STANDARD OF LIVING

What is meant by "standard of living?" It refers to things which make our lives different from lives of primitive societies, i.e., the level of material conveniences and comforts of a "civilized" life. It means being able to purchase manufactured goods—clothing, food, vehicles—rather than needing to satisfy each of these needs personally. As the standard of living rises, less time need be devoted to simple maintenance and upkeep of our bodies and homes. A high standard of living means that improved medical, sanitation, transportation, and protection services are provided for citizens.

The standard of life depends upon the amount of energy per person used in a given community. By the energy available to one person, we mean not only that energy actually under the control of his hand, a switch for the electric heater or TV, but also the *average* of the energy used per citizen (a) in factories which provide materials of comfort, (b) for transportation, and (c) by military forces which use increasingly significant fractions of the country's energy, to name a few other areas of energy use. *So, the more energy a country uses per person in the population, the higher is the material standard of living in that country* (Figure 2.1).

Income, The Standard of Living, and Energy Per Person

It is obvious that the material standard of living you can have depends upon how much money you earn. There are, of course, wide variations in income among the citizens of a particular country. However, in general, we can speak of an *average* income for the citizens of each country (Figure 2.1), and use this to estimate the average level of material comfort in that country.

Energy, Comfort, and Efficiency

The standard of living should not be viewed as referring simply to levels of luxury and leisure. Below a certain degree of "comfort," people cannot work efficiently, e.g., if the weather is too cold or too hot. Therefore, comfort and a fair standard of life are important determinants of effective work in industrialized societies.

This leads to an important idea, which can be referred to as "feedback." The more the energy in a society (and therefore the comfort and feeling of well-being), the better the quality of the work output by people. And the higher the quality of the work force, the greater is their enterprise and achievement. Therefore, more is made by industry, and energy demands keep increasing. But the more energy provided for factories, the greater is their production of cheap goods, in turn increasing the standard of life for that society. Energy allows comfort and, as has been statistically shown, an increased life expectancy. Comfort makes for better work, and this will promote the use of more energy and have a reinforcing effect on the standard of living.

The Divisions of Energy Use

In order to discuss energy alternatives, it is useful to divide energy use into four main divisions:

1. Household energy: heating, cooling, cooking, lighting, appliances, etc.

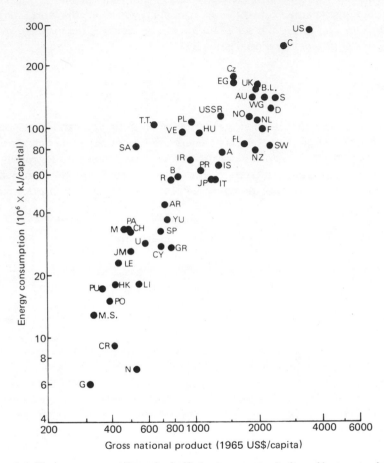

FIGURE 2.1. The average material standard of living in a country (indicated by its rate of energy consumption) is determined by the average income of the citizens of the country. *Key:* US, United States; C, Canada; Cz, Czechoslovakia; EG, East Germany; UK, United Kingdom; B.L., Belgium and Luxemburg; AU, Australia; WG, West Germany; D, Denmark; USSR, U.S.S.R.; NO, Norway, NL, Netherlands; T.T., Trinidad and Tobago; PL, Poland; VE, Venezuela; HU, Hungary; F, France; SA, South Africa; FL, Finland; SW, Switzerland; A, Austria; NZ, New Zealand; IR, Ireland; PR, Puerto Rico; IS, Israel; R, Romania; B, Bulgaria; JP, Japan; IT, Italy; AR, Argentina; YU, Yugoslavia; M, Mexico; PA, Panama; CH, Chile; SP, Spain; U, Uruguay; CY, Cyprus; GR, Greece; JM, Jamaica; LE, Lebanon; PU, Peru; HK, Hong Kong; LI, Libya; PO, Portugal; M.S., Malaysia, Singapore; CR, Costa Rica; N, Nicaragua; G, Guatemala.

2. Transportation energy: driving cars, trucks, planes, trains, and ships, for passenger and commercial transport.
3. Industrial (commercial, business): the energy to run factories, industrial complexes, research and development divisions, etc.
4. Health, education, and welfare activities: large-scale administration, activities concerned with the water supply, with sewage and rubbish disposal, and the energy used by the military forces, hospitals, schools, etc.

Roughly, each of these divisions takes about 25% of the total energy of the country. Each division is discussed in more detail in the following sections.

Household Energy

Heating requires more energy in a household than any other need, about one-quarter of the household total, or $1/4 \times 1/4 = 1/16$ of the whole country's energy. This is why supplying hot water and heat using solar absorbers on the roofs of houses would be a very useful development (Chapter 11). It should reduce by about 6% our use of natural gas.

The next largest energy use is for lighting, which consumes about 15% of the household energy budget. In the U.S., energy used for cooling is not much different than that used for lighting. Household energy use is summarized in Table 2.2.

Transportation Energy

Transportation has become one of the most obvious uses of energy, reminding us more clearly and more frequently of the impending shortage of fossil fuels. With each new price increase of gasoline, usually brought about by a temporary, artificially induced shortage of gasoline at station pumps, we realize our dependence on energy for freedom of movement —commuting to jobs, visiting friends, taking vacations, shopping, etc.. Imagine, we have actually been willing (although certainly not pleased) to wait on lines for *hours*, in all sorts of extremes of weather, to get gas during a shortage! Bribes are paid to local station owners, fights break out, and people have even been murdered during the most recent gasoline shortage in the U.S. in the summer of 1979. It is *crucial* that we realize that although these temporary "shortages" are over once satisfactory price increases are achieved (satisfactory to the large vendors, that is), they reflect the fact that there is a true, underlying fossil fuel crisis—oil *is* running out quickly, and the fuel companies are anxious to use the shortage situation for normal business profit. But beware this "Cry Wolf" game, because the shortage is real!

TABLE 2.2. ANNUAL PRIMARY ENERGY CONSUMPTION FOR A TYPICAL
RESIDENTIAL STRUCTURE IN THE BALTIMORE–WASHINGTON AREA

Component	All-electric house, therms[a]	Minimum-energy house, therms[c]
Heating	480[b]	270(G)
Lights	218	218(E)
Refrigerator/freezer	200	200(E)
Clothes dryer	180	40(E)
Range	120	50(G)
Outside light	90	90(E)
Color TV	54	54(E)
Dishwasher	40	40(E)
Iron	16	16(E)
Clothes washer	11	11(E)
Coffee maker	11	11(E)
Furnace fan	0	43(E)
Miscellaneous	131	131(E)
	1487	1174

[a]One therm $= 10^5$ BTU $= 1.055 \times 10^5$ kJ.
[b]Electric power based on power plant energy consumption of 10,910 BTU/kW-h.
[c](G) stands for gas, (E) stands for electric.

We are not wise in the way we use energy in transportation, particularly in the private car. Having a private car provides more real personal freedom than any other part of our economic and social system, but private cars use much more energy per person per kilometer than many other forms of transportation, particularly trains.

Cars are inefficient users of energy. At a medium speed, they use only about 25% of the total energy contained in the fuel, and the rest of the energy is wasted; some 75% is lost as heat to the surrounding atmosphere. But the main waste of private cars is the carrying space involved—often a large, gas-guzzling car built to hold four to six people is used to carry only one or two.

One approach to reducing the energy we use in transportation is trying to change the technology to one which makes use of the remaining energy more efficiently. Ideas for doing this are being tested. One of them calls for the use of *fuel cells* (see Chapter 12). A fuel cell is an electric power source which converts chemicals such as methanol (methyl alcohol), together with oxygen from the air, into electricity to power a car. This electricity would drive electric motors. Not only is this a more efficient way to use our fossil fuels, it is also cleaner.

Another idea which would help save energy in transportation is to use regenerative braking. Energy is wasted when we accelerate a car up to speed, and then step on the brakes. All of the kinetic energy of the motion of the car is converted to heat energy, which is produced by the friction of the braking system. All of this is *wasted* energy (the heat produced escapes to the surrounding atmosphere), and therefore adds to the inefficiency of the car's energy use. If a mechanism could be designed whereby the force of the car's forward motion could be converted, during braking, to useful work rather than heat, through a reverse fluid drive which would slow the car down but at the same time work a generator, much energy would be saved. We would then be producing electrical energy and storing it in batteries for later use in driving the car. This would be another way, in addition to the use of fuel cells, to save energy in transportation.

Industrial and Commercial Energy

Consider the manufacture of a car. We start out with iron ore, which is found in the earth. We extract it from the earth, and subject it to a process in furnaces in which we mix the iron ore with coke (a form of carbon). These substances react and produce iron, releasing wasted energy in the form of heat into the air, and polluting the skies with carbon monoxide and carbon dioxide.

Carbon is then added to the iron, and the iron and carbon react to form steel, at a temperature of about 1650 °C. This metal is cooled to solid steel, which is a malleable metal, easier to bend and more difficult to break than iron. This flexibility is what makes steel more valuable in industry than iron.

Then, the sheets of steel are put into machines which, using large amounts of energy, bend it into the shape of the car body. At the same time, other machines are making engine blocks, pounding steel into the required shapes of pistons, cylinders, transmissions, etc.

Energy in industry is used in several forms. Heat is one form. It is given off and wasted during the process of melting iron ore at 1650 °C. Another form, electricity, runs some of the machines, and factories are heated and lighted by it.

Industry consumes about one-quarter of all the energy we use. Along with energy-using transportation and household heat and light, etc., it establishes our standard of life. The level of comfort that can be achieved within a given society will be determined by the price of energy in that society, as well as by its members' ability to attain a certain income within the social system of the country.

FIGURE 2.2. In addition to polluting our atmosphere when used as a fuel, oil also pollutes our waterways when accidental spills occur. Here, Dr. Ananda Chakrabaty is studying a microbe which can eat oil faster than any other known organism, and could help minimize oil spill damage.

ONE (AVOIDABLE) NEGATIVE ASPECT OF A HIGH-ENERGY LIFE: POLLUTION

Until recently, in the past decade, it did not occur to most people that a price would be paid for industrialization of an area. In general, industrialization means new jobs, growth of towns, needs for new services, supermarkets, doctors, etc. In other words, industry brings money, growth, and many new opportunities. However, we have finally realized that there is a penalty associated with this growth, and that penalty is pollution of our air, our water (see Figure 2.2), and our land.

It is becoming apparent, however, that it is not necessary that a high-energy life leave a dirty trail. It depends on what the source of energy is. Coal? Oil? Natural gas? Or is it the atom? Each of these sources has a different degree and type of polluting power. Coal, for example, is dirty, because it gives off a lot of grit and coal dust. At the same time, because coal contains sulfur, we get sulfur dioxide in the atmosphere when coal is burned in a furnace to give electricity, and this is what sometimes causes smarting of the eyes (see Figure 2.3).

Oil is dirty, too. The products which come out of the exhaust pipes of cars contain poisonous materials (Figure 2.4), which are spewed out into the atmosphere we breathe.

Natural gas, or methane, is a cleaner fuel than is oil. However, it does cause the ejection of carbon dioxide into the atmosphere, and this does not all get converted by photosynthesis to oxygen and carbohydrates, so the carbon dioxide level in our atmosphere is gradually building up.

Atomic (fission) reactions produce poisonous and dangerous materials and are discussed in Chapters 6 and 7.

There seem to be many illogical aspects of our present energy system, and that is one of the reasons for writing this book. Why tolerate dirt-pro-

FIGURE 2.3. In our present energy system, high levels of energy consumption are accompanied by high levels of pollution in our atmosphere. (Left) A clear day with essentially no air pollution. (Middle) Smog is trapped by a temperature inversion layer at ~90 m (300 ft) above the ground. (Right) Smog engulfs the Civic Center area on a day when the inversion layer is ~460 m (1500 ft) above the surface.

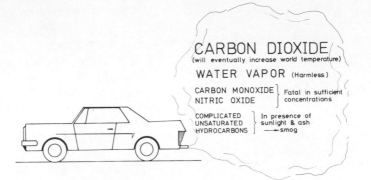

CARBON DIOXIDE
(will eventually increase world temperature)

WATER VAPOR (Harmless)

CARBON MONOXIDE] Fatal in sufficient
NITRIC OXIDE] concentrations

COMPLICATED] In presence of
UNSATURATED } sunlight & ash
HYDROCARBONS] ——smog

FIGURE 2.4. Although often barely visible, automobile exhaust contains poisonous substances and causes much of our harmful smog.

ducing energy sources? Our future energy sources should be clean ones. Then, the advantages of a high-energy life will not be spoiled by air and water pollution, which is due mostly to our use of fossil fuel energy sources—oil, coal, and natural gas.

Another Negative Aspect of the Present Energy System: The Coming Exhaustion of Our Present Sources

A major difference between modern people and preindustrial people arises partly from the fact that we learned how to use energy. In the remote past, fire was first used by burning wood, then coal, and much later (not until 1859), oil. Later still, we began to use natural gas, then, quite recently, the atom, and now there is a prospect of our being able to utilize solar energy on a large scale. We built up our technology, and thereby increased our standard of life, largely by using the coal and oil, not realizing that the supply of these fuels would last for a mere century and a half.

When man first began to use coal and oil, he was like a person who had come across what seemed to be an unlimited supply of money. It was as though we had found a limitless source which we could go on using, without ever seeing the bottom of the barrel. We began to use coal as a fuel in the 18th century, and extensively in the 19th, when oil was also discovered as an energy source. We developed the use of oil in the 20th century, and then also began to use natural gas. The use of atomic energy is a post-1950 development and hardly counts as yet, because it gives (in the 1970s) only some 1–2% of energy in a technologically advanced country such as the U.S. or France, and makes as yet a negligible contribution to the energy supply in most countries.

Now we find that the supply of oil and natural gas is going to decline in the *near future*. Then we shall not have enough energy left to run the equipment which gave us a reasonably comfortable life, if we do not get the machinery of an alternate energy source built in time.

Think of what will happen when we do not have energy from oil, natural gas, and coal, whilst at the same time the world's population is rapidly increasing (Figure 2.5).

Suppose that a person inherits a sizable sum of money through the death of a relative. Suddenly she realizes that she has a lot of money and what seems to be unlimited purchasing power. If she is a spontaneous, short-sighted person, she will simply continue spending it until one day her bank account is down to the pre-inheritance level. Then she will have the very difficult task of returning to her previous, less extravagant life style. If she is more levelheaded, she may spend her inheritance more slowly. How-

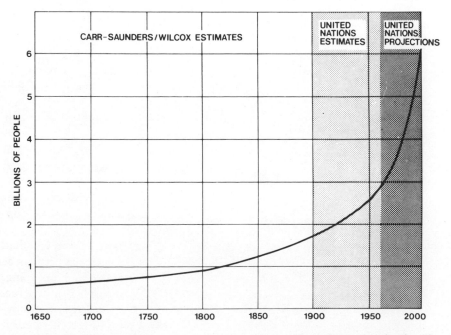

FIGURE 2.5. The rate of population increase has risen dramatically in the past few decades. The achievement of zero population growth, or perhaps negative population growth, should be one of our main objectives in solving the ever-increasing energy crisis.

ever, it will still run out someday. Finally, if she is clever and plans for long-term periods, she will invest her money so that it will bring her money from other sources, other than the single inheritance. In other words, she has turned her nonrenewable inheritance into a renewable resource of capital which can be reinvested indefinitely to bring her the money she needs to maintain her new, higher lifestyle.

3

How Do We Get
Our Energy Right Now?

INTRODUCTION

It is worthwhile, at this point, to restate what energy is, a question that we discussed in Chapter 1. We have already seen that, formally, energy is a force pushing through a distance, whereby we say "work is done." We can feel the essence of motion (or kinetic energy) in the concept of the *movement* of a body, and in the push necessary to make it attain its velocity. We have also seen that bodies can have potential energy as a result of their position (e.g., a weight raised to a height above the floor), and that substances can have energy stored in the chemical bonds of the molecules which comprise them, energy which may be released as heat when the bond is broken.

In this chapter, we are going to discuss how we get our energy *at present*, so that we may better comprehend the energy disaster developing as our present energy sources are depleted, and what requirements our energy alternatives must satisfy in order to not only adequately replace our present sources, but to be better sources which will not give us the same problems at some future time.

A SUMMARY OF OUR PRESENT ENERGY SOURCES

There are two main ways whereby we get our energy. The first of these is via heat, and the second, via electricity (which, in fact, is obtained from heat energy).

The most common way we get energy is to combine oil or natural gas (hydrocarbons—substances consisting of hydrogen and carbon), obtained from deposits in the earth, with oxygen from the air to produce heat. We call this "burning". One type of burning is called "external combustion," an example of which is a gas kitchen stove. Here, natural gas enters through a pipe into a burner. When the methane (natural gas) is emitted through holes in the burner, it combines with oxygen in the air and burns, producing a flame and heat energy. Carbon dioxide and water are formed, but they are colorless and odorless and harmlessly diffuse into the air.

Another way of using the heat energy produced in a burning chemical reaction is with "internal combustion" (Figure 3.1). This is what happens within the cylinder of a car, for instance. The cylinder contains a piston, and when the piston moves, it causes energy from an electrochemical battery in the car to give a spark. The gas inside the cylinder is a mixture of gasoline vapor (hydrocarbon vapor, like decane or octane) and oxygen from the air. The spark initiates combustion, and it takes place so fast in the enclosed space that it is actually an explosion. When the explosion occurs inside the cylinder, the water vapor and carbon dioxide produced in the reaction build up pressure because of the heat released in the reaction, and push against the piston. The *energy* produced is equal to the force of the explosion times the distance the piston moves. The piston is connected by a rod and shaft to a rotating device called the transmission. This then rotates, whereafter, through more gears, it turns the wheels of the car (shown in Figure 3.1). When we actually *drive* a car, the explosion and the piston-forcing effect occur many times per second, and the car engine typically has 4, 6, or 8 cylinders, with each piston connected to the turning shaft. The rotation of the transmission then becomes smooth, though at low speeds in high gear we can still sometimes just feel the throbbing caused by the explosion (a chemical reaction that releases potential energy; see Chapter 1). Carbon dioxide and water vapor that are formed exit out of the car's exhaust pipe.

Some other chemicals, which are *not* harmless, come out of the exhaust pipe too, and this is why we are able to *see* the exhaust (carbon dioxide and water vapor are invisible). These other substances are about 1% of the total products and are formed in side reactions, rather than in the main combustion reaction. These other reactions are an important problem because they form dirty, unhealthy smog in the air.

Making Electricity

In creating electricity, heat from chemical reactions is again used (the combustion of oil or natural gas, or perhaps coal, with oxygen, to form carbon dioxide and water). The flame heats a boiler, water is boiled, and steam from the boiling water spins turbine blades. The turbine rotates an electric generator between magnets which produces electricity in the wires coiled around the generator.

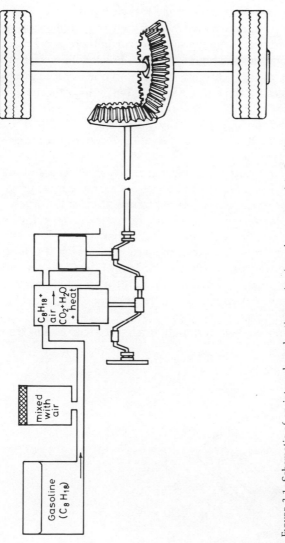

FIGURE 3.1. Schematic of an internal combustion engine (not drawn to scale). Gasoline is mixed with air and exploded inside the cylinder by a spark. The heat causes gas to expand, which pushes the cylinder, giving mechanical energy to the drive train which transmits it to the wheels.

Getting electricity from a generator depends on a concept called "magnetic lines of force." These are lines of magnetism which we imagine to run between the end pieces of a magnet (magnetic poles). When we turn the "arms" of a generator, these lines of force get broken and "induce" electricity to flow through the wires of the "armature" (Figure 3.2). This electricity flows down (i.e., an electron stream travels inside) the wires from the armature and out through the wires of the electricity-generating plant.

When first produced, the voltage and current which are made at a central generating plant are too high to use in a household or factory. We send electricity at high voltages through wires from one town to another, if necessary, and it enters a "transformer substation." At such a station electricity is "conditioned," which means it is reduced to a voltage which is less likely to cause shock and damage (generally 110 in the U.S., 220 in countries of British origin). In that form, it is suitable for delivery to houses and factories.

Thus, when we switch on an electric motor in a factory or an air conditioner at home it is *chemical* energy we are using. The energy is coming from the stored potential energy in the bonds of oil and oxygen molecules, which have been converted to kinetic energy of the gases in the heat engine, and then into electricity by the generator driven by the heat engine.

Electricity's main advantage is the ease with which it can be sent over distances (*short* distances, of 100 miles or so). Electricity also is turned on and off simply by means of a switch. At home it's easy to switch on an electric heater rather than spending thirty minutes building a coal fire as previous generations had to do.

It should be noted that, although much electricity comes from coal and

FIGURE 3.2. How an electricity generator works.

FIGURE 3.3. The hydroelectric generating units of the Wheeler Dam on the Tennessee River. Total generating capacity is 356 MWe in 11 units. (Courtesy of Tennessee Valley Authority.)

oil, quite a bit also comes from hydropower (Figure 3.3). See also Figure 3.4, which is a schematic of a hydroelectric plant.

FOSSIL FUELS

In all of these cases of energy production, the energy comes initially from fossil fuels (Figs. 3.5–3.7). During combustion, we combine fossil fuels with oxygen from the air (we use air without trying to separate the oxygen), ignite the mixture, and harness either the forces of expanding gases from an internal combustion engine, or heat if we used external combustion. These forces can work pistons or turn generators to produce electricity.

But what are fossil fuels? They are coal, oil, and natural gas. They are organic substances buried in the earth many hundreds of millions of years ago, which arose from the decay of plants, and composed mostly of carbon and hydrogen.

Coal is found as lumpy deposits in the ground, buried for the most part a few thousand feet deep. We have to make mines in the earth to get to the coal seams. Removing the coal from the mine to the surface (Figure 3.8) is messy and difficult. There are few fully automatic devices for getting coal

Figure 3.4. Schematic of a hydroelectric plant.

from where it has sat for hundreds of millions of years up to the earth's surface. People are lowered into the depths and use drills to break the stuff out lump by lump.

Oil is usually found in pooled deposits underground. When we are lucky enough to find an oil-bearing underground formation, we drill to the level where the oil is thought to be. Sometimes it spurts up to the wellhead under high pressure. It may catch fire and burn vigorously until we get it under control. A cap is then put on the spurting hole, and later, when the exuberant original pressure has died down, pumping up of the oil is begun. At refineries, crude oil is separated into different parts, each having a different volatility and weight. Later, the refined products are sent to factories and gas stations.

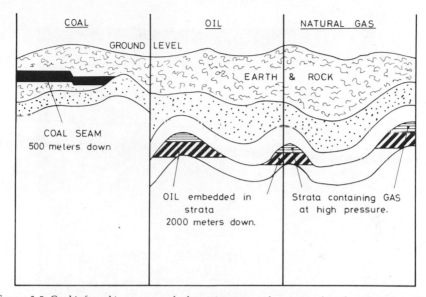

FIGURE 3.5. Coal is found in seams, and when mines are sunk, men work at the mine face to dig out coal. Oil is embedded in strata called oil fields. When a field is discovered, oil spurts out of the ground at first, and later has to be pumped out. Strata in the earth also sometimes contain natural gas at high pressure. When discovered, it flows out and is connected to a pipeline system.

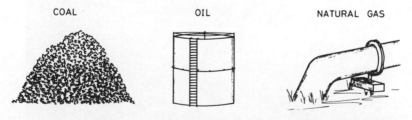

FIGURE 3.6. Coal is stored in heaps on the ground after it is dug out of the ground. Oil is stored in tanks after it is pumped out of the ground. Natural gas is left in the ground, with an outlet system arranged so that the gas passes through purifiers (not shown), and is then piped to the towns.

FIGURE 3.7. The Moomba site of natural gas collection in the South Australian Desert.

FIGURE 3.8. Strip-mining operation.

Lastly, there is natural gas, or methane. We find what we call "fields" of this, which are often 2000 or 3000 meters deep. A hole is drilled into the earth, and as the drill goes deeper, instruments indicate when its head is in contact with gas. The gas is under pressure and rushes up out of the hole we made. A cap is put over the hole to keep the gas from escaping before it can be piped. The collected gas is pumped to a refinery where unwanted substances are removed from the gas. Then the refined gas is sent to surrounding towns for households and industrial centers. Sometimes it is distributed in pipes over several hundred, or even several thousand, kilometers (Figure 3.9).

We can say that coal, oil, and natural gas are our solid, liquid, and gaseous fossil fuels, respectively. Figure 3.10, an oil-fueled electricity generating plant, is of interest.

FIGURE 3.9. The Alaskan crude oil pipeline during its construction.

How We Get Our Energy in Industry

This is a more complex situation than that of energy for households and transportation. The reason is that there are many different uses of energy in industry. It depends on the industry and what is being manufactured. Steel? Pepper pots? Dictaphones?

We can illustrate with two examples. First, many industrial machines have something rotating—gears which drive cutting edges or which form shapes out of large chunks of metal. Here, often, we use an electric motor to drive a rotating shaft. Of course, in most cases the electricity to do this comes from coal or oil, as explained previously. Sometimes, however, machines are driven by natural gas. For example, steam hammers are raised in a shaft by steam pressure, created from natural gas, which then drop down on a metal to shape it. The steam hammer is raised by the energy from steam pressure; at full height, this mechanical energy is in the form of stored potential energy and is used up in the downstroke of the hammer to beat the metal into shape. Here, the energy again has to come from coal, natural gas, or oil, which reacts with the oxygen of the air to supply heat energy.

This heats water, generates steam, and lifts the hammer which performs work.

Lastly, many machines run on natural gas (Figure 3.11), which is used in the same way as described above: The chemical process of burning gas gives off energy in the form of heat, causing expansion inside the cylinders and pushing the pistons which drive the gears and turn the wheels.

What is the summary of all this? It is *simple*. The same thing always happens. The chemical bonds in the hydrocarbon fuel and the oxygen are rearranged to form the bonds in the carbon dioxide and water produced by the reaction. Because more energy is contained in hydrocarbon and oxygen molecules than in carbon dioxide and water molecules, the remaining energy is given off as heat. This heat speeds up the motion of the molecules of the product gas, causing the gas to expand, which forces the piston to pass through a distance. Force passing through a distance means work is being done, and energy is being used. Stored potential energy of the fuels has been converted to useful work energy.

FIGURE 3.10. The 991-MW, modern oil-fueled electricity generating plant at Huntington Beach, California.

FIGURE 3.11. This typical modern electricity-generating facility supplies electricity, obtained from burning natural gas, to about a million people. The small size of the facility is especially noteworthy.

PROBLEMS OF THE NEAR ENERGY FUTURE

If we go back far into the past, the potential energy stored in the bonds of the fossil fuel molecules came from the sun, through photosynthesis, as explained earlier. In the future, we face two problems with respect to energy. The most important one is deciding what the new sources of energy are going to be when the fossil fuels are gone, which may take place during the lifetime of many of the readers of this book.

The second problem is how to use this energy so that it is clean and does not cause air pollution and smog (as do some products formed from

fossil fuels)—to say nothing of the dangers of cancer from atomic fission reactors now in use (more about this later, in Chapter 7). Thus, we want to
know whether we shall produce energy in the form of electricity, which is
clean, or use a new idea, such as putting it into the gas hydrogen, which can
be stored in tanks and transported through pipes. Alternatively, perhaps we
could transform it into a convenient liquid like methyl alcohol, or methanol, which could be stored in tanks, like gasoline (more about storage in
Chapter 16).

How Energy is Transported Now

We have seen that 98% of our energy comes at present from fossil
fuels. These have given us our present industrialized civilization, but will be
eventually exhausted. We must see to it that new, safe energy sources are
built, or soon our standard of life will break down. Were this breakdown to
occur, the populations in the towns could not be supported, because life in
cities with transportation, food and water distribution, heating and lighting,
and factories, is very dependent on a daily supply of large amounts of energy.

For this same reason, it is clear that energy *does* have to be sent over
great distances. It is rare for people to build a town on top of a coal mine,
near an oil field, or near a natural-gas field. Hence, *transport* of energy is an
important aspect of the energy arrangements of the future.

First, the transportation of coal. Coal is sent in solid form, mostly in
trains after it has been hacked out of the earth and brought to the surface.
Hauling coal around the country in trains has not changed much in 100
years.

Oil is transported through pipes and these can be very long, some of
them 4000–5000 kilometers in length. There are pumping stations to continue to push the oil onwards, for example, from the center of Saudi Arabia
to the great oil depot at Russ-El-Tunurra on the Persian Gulf, where it is
put into giant tankers (up to 500,000 tons in loaded weight). These tankers
then travel around the world, for example, to Antwerp, in Belgium, to supply part of Europe, or to Yokohomo, to supply part of Japan, or to Chester,
Pennsylvania, to supply part of the U.S.

With natural gas there is an easier situation, because gases flow easily
through pipes with less pumping energy needed than with oil. Hence, natural gas is pushed through pipes over long distances, and the pipes can be
much smaller and still carry enough natural gas to support a big city. A
town of a million people can receive all the energy it needs right now from a
natural-gas pipeline some 60 centimeters in diameter.

Most of us have seen how electricity is carried from town to town
through wire grids of high-voltage cables fixed onto towers.

Our Finite Energy Sources

We must stress that our present energy sources, the fossil fuels, are a one-time gift from the solar radiation of the past millions of years to present man. There is just so much which was formed through the millions of years in the past, and no more. It is like a single stack of money. Of course, it is conceivable that, in hundreds of millions of years into the future, the world will go through another cycle, and perhaps the cycle of growth and transformation will once again happen, forming a new supply of fossil fuels. But it is more probable that we shall cut down most of the forests and use up our wood to burn and supply energy, so that little will be left with which to form a second coal–oil age.

That is why we have to be clever. We have to create in a short time, in just about two generations, an *international array of energy-producing factories and connected pipelines,* to use the inexhaustible sources of clean energy which we shall discuss in this book.

4

Shall We Run Out of Energy in Our Time?

PRESENT FUELS ARE LIMITED

We have seen in the last chapter that the fossil fuel energy sources which we have been using will be exhausted. The amounts of all of these sources initially, before we started to use them, are shown in Table 4.1.

Because we are burning more of these fuels each year, yet have only a given, limited amount of fuels, there will come a time when fuels such as oil, natural gas, and coal will be so scarce that their cost will be greater than that of the alternative energy possibilities. In this chapter, we will estimate how much time we have left before the price rise is such as to make imperative the availability of alternate fuels. Then we can calculate, for a certain given rate of growth of the economy, how long it will be before we must have the new, abundant, clean energy sources from the sun and the atom ready to replace the fossil fuels.

HOW DO WE ESTIMATE THE FOSSIL FUELS LEFT?

To estimate when the coal, oil, and natural gas in the world will exhaust, we must consider the total resources left, the rate of consumption at the present, and extrapolate to rate of consumption in the future, taking into account the yearly rate increases due to population growth, economic growth, and resultant rise in demand. For example, in 1960, the amount of known reserves of oil in the U.S. was 165 billion barrels, and the use rate about 3 billion barrels for the year. So it seemed that we would be able to go on using oil as a fuel for 55 years from U.S. sources alone if the rate of consumption held constant. However, during the last few decades, the rate of use of our resources has been increasing, i.e., more consumer products are

TABLE 4.1. ENERGY CONTENT OF THE WORLD'S INITIAL
SUPPLY OF RECOVERABLE FOSSIL FUELS

Fuel	Quantity
Coal and lignite	7.6 trillion tons
Petroleum liquids	272 billion tons
Natural gas	10 trillion ft^3
Tar–sand oil	41 billion tons
Shale oil	26 billion tons

sold each year than the year before, the energy consumption hence increases each year.

The complication of the rate of consumption increases each year is illustrated in the following tale.

The French Farmer

Once upon a time, there was a French farmer who had a pond. On this pond there grew lilies. The farmer was slightly worried about this because he did not want lilies to cover his pond, which contained large fish which the farmer would sometimes catch and eat for breakfast. However, being a rather easygoing and relaxed sort of fellow, he said to himself: "I'm not going to bother about those darned lilies until they have covered half the pond, and, anyway, it will probably never happen."

Now, the farmer had a son who was of a more studious, serious, and observant nature. He watched the lilies grow day by day, and recorded what he saw in one of several notebooks he carried.

The farmer continued happily along, making a profit by selling his French sheep and eating his occasional breakfast of fish from his pond. But his son, while puzzling over his books one day, came to a startling conclusion—everyday, the number of lilies on the pond doubled! He told his father about this, and they went out to look at the pond. To the farmer's surprise, it was half-covered with lilies. "But it couldn't have been more than a quarter covered with the confounded flowers yesterday!" he exclaimed in dismay, not very quick to catch on. "I suppose I'll have to clean the pond this week," he grumbled. (See Fig. 4.1.)

Of course, the son realized that his father did not comprehend that the pond would be *completely covered tomorrow*. When he finally succeeded in getting this point across to his father, the situation was unfortunately out of control.

FIGURE 4.1. The French farmer.

RESOURCES EXHAUST SUDDENLY

The story of the French farmer exemplifies the suddenness with which resource exhaustion can expand beyond control. The rate of consumption of our resources is growing each year. Figure 4.2 shows the rate of consumption of oil in the United States, and how it increases with time. The rate of consumption of hydrocarbon fuels increases each year (Figure 4.3). However, after a certain date, the rate of production will go rapidly down, while the demand continues to rise.

At the beginning of the use of a resource, there seems to be a large, perhaps almost infinite, amount in the earth. At first, it is used up slowly, and the amount in the ground hardly seems to change. This corresponds to

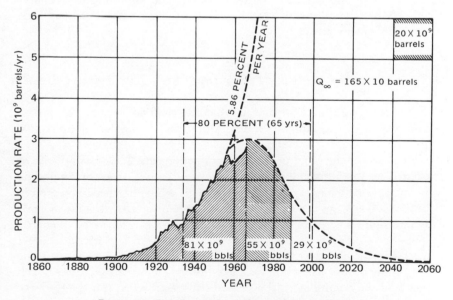

FIGURE 4.2. Complete cycle of U.S. crude-oil production.

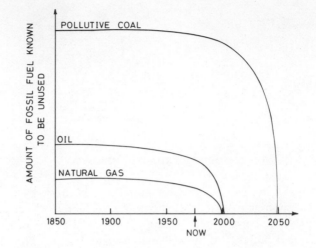

FIGURE 4.3. Exhaustion of resources is unnoticeable until shortly before the end.

the initial part of the graph shown in Figure 4.3. After a number of years, the amount we use in one year is a significant fraction of the total still left (say, 10%), and we begin to notice a decrease in the total amount still available, as shown by the latter part of Figure 4.3. A few years after we have begun to notice the decrease of the resource, the amount still available quickly *crashes* down toward zero, i.e., the resource *suddenly* exhausts (cf. the French farmer's story).

The Date of the Maximum Production Rate—The Effective End

How can we estimate the date at which we shall no longer be able to go on pumping more oil, pumping more natural gas, or digging up more coal to meet our expanding demand?

First, assume the point of view of production *rate*. If we have plenty of a resource, there will be an increased dependence on its use, corresponding to the increase·of the energy demand per year, which has been going on in affluent countries, especially in the Western world, for decades. This is shown in Figure 4.4. At first, energy production rate (and therefore energy consumption) increases by the same amount each year, but after a while, it climbs more steeply each year than the previous year. Thus, in terms of the rate of production of the resource, a maximum rate is reached, after which the needs of the community are likely to still be growing, but the resource reserves will be declining, and will not be able to supply all of their need. The production rate will, in fact, go down, as shown in Figure 4.5.

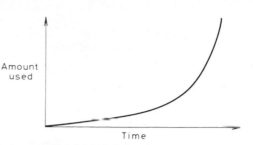

FIGURE 4.4. Productivity increases with
time—exponentially.

The time at which the production rate passes through a maximum is the
effective date at which we need to replace the exhausting resources from
some other resource, e.g., replace oil with wind or solar power, using hydrogen or methanol as an intermediate fuel (Chapter 15). A compromise
would be to stop growth of both population and living standard. The best
solution would be, of course, to stop the population growth, but not the
energy growth.

When Will the Maximum Production Rate Be Reached?

To calculate the year in which the maximum rate of production of a resource will occur, we must know what the total amount of the resource
available is. Since new discoveries are still being made, we can only roughly
estimate these totals.

We will also have to know the rate of increase of the use of the resource
in the future. Will this increase, say, at 3 or 5 times each year? Predictions
of how much a community will increase its use of a type of energy in the future (see Fig. 4.6) will always involve some uncertainty because expansion
rates may speed up—*but they may also slow down.* Hence, the prediction
of when we may run out of oil or coal can only be made by saying that *if* we
go on growing as we have been growing, then the production rate of, e.g., oil

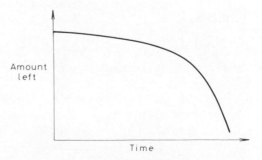

FIGURE 4.5. If productivity, and hence
consumption of fuels, increases with
time, the amount of fuel left will decrease correspondingly.

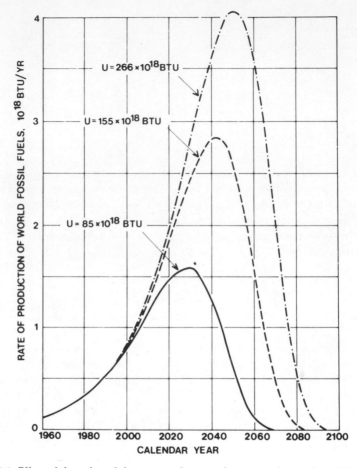

FIGURE 4.6. Effect of the value of the resource base on the projected rate of world fossil fuel yields. (Results from Elliot and Turner.)

will pass through a maximum around a certain year. This year turns out to have been in the 1970s for U.S. oil (Fig. 4.2).

Actually, this method of estimating future situations is very helpful, because it lets us know if indeed we must slow down our consumption rates to allow time for alternate energy sources to be *fully* developed, and if so, by how much.

In addition to the causes for economic growth and increasing energy demands already mentioned, there is another, more subtle and indirect, reason for growth. There is an increasing need to *create* jobs for people, whether or not those jobs are directly tied into the needs of society. This is because our industrialized world is being increasingly run by machines, displacing many workers and requiring only one, or perhaps no, human workers to run them. At the same time, while the demand for workers is decreasing, the Western world retains to a fair (if decreasing) degree the work ethic: unemployment is still equated with laziness, dishonor, and failure. We make the condition of unemployment, even when uninvited, very miserable, by allowing unemployed people to have limited incomes, just enough to keep them well. Of course, this problem is compounded by population increases and migrations.

Hence as machines displace people, either unemployment rises by several per cent a year, until the point of unacceptability, or the community grows, and more suburbs open up to give rise to more vacancies for the kinds of jobs still done by people (the so-called "white collar world"). But this expansion increases our need for energy per person per year, and hastens fossil fuel exhaustion.

The Effect on the Exhaustion Date of Rising Living Standards and Population Growth

Although often taken for granted as being the natural progression, we might pause to ask why it is that there *is* a rising use of energy, meaning an increase in production rate of oil and other energy resources. The answer is simple: the increase arises in two ways. One cause is the growth of the world's population. Since good medical services have become widely available, the majority of babies that are born will live. Meanwhile, the average lifespan keeps increasing. So we always have an increase in the number of mouths to feed, bodies to clothe and transport, etc. *Maintaining* our standard of living, therefore, requires more energy each year as more people are making the same increasing demands.

There is another reason for the growth of the use of energy: living standards in the industrialized countries of the world have been going up. Rising living standards mean more energy used per person.

It is this increase in living standards—an increase in the amount of energy used per person, coupled with an increase in the number of people in the world—which brings the increase in the need for energy and which plays a part in the exhaustion of coal, oil, and natural gas at an earlier date than that at which we thought it would occur decades ago (Fig. 4.7).

How the Rate of Production of a Resource Will Vary with Time

The rate of production will depend on demand and supply. Thus, the rate at which a resource will be produced reflects the pull of the market, but

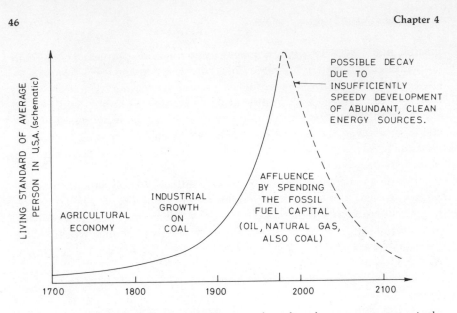

FIGURE 4.7. Living standards for the average person depend on the energy per person in the community. Oil and natural gas fuels were discovered in the nineteenth century, and our civilization has been riding high on the crest of technology ever since. But oil and gas will soon be exhausted. Will machinery be ready for the collection of energy from inexhaustible sources, or will living standards undergo disastrous decay?

there is also the question of *how much* the total supply is. Qualitatively, it's clear that when the demand is small, relative to the supply, production rate will be independent of the total quantity of the resource. Eventually, however, there will come a time when there is no more resource left. As that time approaches, there will be greater and greater limitations on the rate of production of the resource—the supply of energy from coal, for example—and then, finally, the rate of production will have to fall to zero.

The question of demand is more difficult to discuss. The rate at which a resource will be in demand is likely to be a fraction of the amount of the production which has already occurred. This is the essence of the growth concept: the more one had, the more one wants.

The Critical Year of the Maximum

Initially, supply is small, and demand is a fraction of this and is also small. As supply increases, demand increases, remaining some fraction of the supply. This, in turn, is some portion of the total amount of the resource. However, at some time (for supplies which are limited and nonrenewable), the supply becomes a significant fraction of the total resource, which is being depleted. The demand continues to increase because society has become used to a certain standard of living with certain periodic increases, and populations continue growing, too. Switching to another resource would take time and money, and people would not get the same increases they've come to expect each year.

So, resources fall, supply falls, but demand *will not easily fall*. That would mean asking people to accept a lower standard of living, which is unlikely to be a popular request. Thus, the rate of production with respect to time is that of an increase while supply and demand are still small relative to total resource. Then, we pass through a maximum, when supply and demand are just balanced by remaining resources. Finally, supply and demand decline to zero, when the end of the total resource is coming into view.

From the point of view of energy resources, the year of the maximum is the critical year. After that year, supply can no longer keep up with demand. Either demand has to fall (and living standards, and eventually life as we know it), or there has to be *ready and built*, another, preferably inexhaustible, supply from a new resource.

THE EFFECT OF CHANGE IN PRICE OF A RESOURCE ON ITS USE RATE

Until now, we have discussed the factors (total amount in earth, rate of use, increased demand each year) which determine when we shall pass through the maximum use-up rate (see Figures 4.2 and 4.6), without considering the effects of resource exhaustion on the price of the resource. As the end of the resource is approached, the resource will go up in price. Partly, this rise is due to "commercial" reasons—sellers know that they can raise the price of the energy because the public needs it (almost like it needs air). But there may also be an increase in cost of digging up or pumping further amounts of the resource, i.e., we are paying for the increased difficulty in obtaining the resource due to its extinction. For example, at first, getting coal by strip mining is easily (hence cheap), but later mines will have to be dug deeper, and this makes the coal more expensive because one needs a lot of energy and manpower and machines (all of which are costly) to reach the coal.

A sufficient rise in price will cut down the use of the resource; people will not be able to afford as much of it. Living standards will fall—the car(s) will not be used as frequently, houses heated and cooled less, etc.—unless we can replace the exhausting resource (e.g., oil) with another source of energy in time.

There is a favorable aspect to this rise in price and subsequent fall in use rate. It will stretch the resource out and make it last longer. It will give us more time in which to do the research and development of the inexhaustible sources upon which so much depends.

How an Estimate of 1500 Years of Coal
Can Be Revised to Less than 50 Years

Some of the points made above can be illustrated by discussing a statement which used to be made about the amount of coal in the U.S. It has been said that the U.S. has 1500 years of coal. This type of statement allows one to relax, and not realize the immediacy and urgency of the need for development work on the machinery for extraction of energy from inexhaustible sources. However, it should be clear from what has been said above that the statement may be misleading, because it was made on the basis of taking the total amount of coal in the ground, regardless of its accessibility, in the U.S. and dividing it by the rate of consumption at the time the statement was made (ignoring changes in this rate).

Let us note some of the reasons why the statement is overly optimistic, and even unrealistic. First, in the U.S., only some one-quarter of the coal in the ground will be available. Taking into account this accessibility factor, only one-quarter of the 1500 years of coal will be available, and this is only some three hundred and seventy-five years.

The second factor, already referred to, is the growth of the economy. To find out how long the coal would last, we have to choose a given year and assume (and it is not a very likely assumption) that from that date on, there will be a constant rate of use of the resource, with no further increases.

We could choose, for example, the year 2000. From the graph which has been plotted for the use of fossil fuel per year (Figure 4.6), by the year 2000 the energy demand should be three times greater than what it was in the mid-1970s, assuming that the past growth trends continue. Hence, focusing out attention on the year 2000, the amount of coal on this basis will last only some 375/3 = 125 years further. This is a sobering reduction from the 1500 years of the former prediction, perhaps, but still over a century away.

However, there is one more modification to be taken into account, indeed the biggest consideration of all from the standpoint of our present energy situation. That is the effect of the decrease of availability of oil and natural gas on the rate of consumption of coal, if this is the only energy alternative available.

When we make a production-rate-versus-time plot, we use the figures of past production rates each year as the basis of our curve. However, coal has not been a source of more than about 15–20% of the energy supply in recent times. If we are going to have to switch over from oil and natural gas when they exhaust, to coal, (neglecting contributions from solar and atomic sources, since these will still be small in 2000), then from about 2000, with coal as the only source of energy, assuming the total demand for energy continues to grow at the same rate as in the past, the number of years of coal left is calculated to be about 30 years worth after the year 2000.

The assumptions made here are certainly very much oversimplified. Obviously, we shall not just stop using oil and natural gas and switch to coal completely in the year 2000. Atomic energy is likely to contribute perhaps 20% of our energy by 2000. Solar energy will be in use by 2000, probably largely for housing, but also with a few massive solar plants already in operation. But what these estimates and figures stress is that we shall not have an *abundance* of coal to turn to after the oil runs out, if we consider availability, price, expansion, and increased use-rate. The supply of coal could not last for the next 1500 years as once believed. At best, coal could be an auxiliary fuel to keep us going until the research on alternative energy sources is done, and the switchover to these new systems completed.

When Shall We Run Out of Oil and Natural Gas?

Before we predict a date for exhaustion, we need a definition of the word "exhaustion." Say we call it the date at which we reach the maximum of the use of the resource, as discussed above. What will be the rate of increase of need in the future? If the price of the commodity (coal, for example) goes on increasing in terms of dollars of constant value,* will this not cut down on the growth of the rate of consumption and stretch out the time over which the resource can be used?

For example, if the expansion rate of the economies of the Western world countries, particularly the United States, follows the path of recent years, and if the price of oil does not rise more than four times what it was in 1973, then the oil will run out in the Western world countries between 1990 and 1995. Natural gas will exhaust at about the same time.

Notice that we have not said for what specific countries these state-

---

*Simplistically, inflation is the decrease of the purchasing power of the currency. Everything gets higher priced during inflation, but in a sense, that doesn't necessarily make it more expensive because it may be that the average income, after taxes have been paid, increases faster than the price of goods. So, to know whether things will become more expensive in a *real* sense, we have to price them in the constant dollars of a given year.

ments apply, but only "for the Western world." This is because there is an excellent international network of suppliers and tankers carrying oil to all parts of the Western world. Most of the oil which supplies the economics of the Western European countries (e.g., France, England, Germany, Italy) comes from countries such as Saudi Arabia, Iran, Kuwait, and Venezuela. China and Russia (and other countries having planned socialist economics) will be in a different situation. For one thing, the standard of life (and hence, the use-up rate of fuel supplies in those countries) is lower than in countries where the governments are oriented to the "enjoy it now and let tomorrow take care of itself" philosophy. Also, if they have been spending at a lower rate, they will have more left over—more oil and natural gas reserves—than the Western world countries. Their energy difficulties will come later than ours.

In the United States, about half of the oil comes from domestic U.S. sources, mostly from oil wells in Texas and Oklahoma, as well as Alaska, and part of the remainder from Venezuela. In the 1970s, however, a disquietingly increasing fraction of U.S. oil reserves came from the only large and easily accessible world oil resources remaining, those of the Middle East, which are in possession of Arab nations.

RUNNING OUT OF COAL, TOO

We have stated that there were illusions in the past about how long coal would last. Many countries have large amounts under the ground, in terms of tons, but the point is that the expansion rate of the economies, and the need to replace oil with coal, will use up the coal about 10–20 times faster than had initially been expected. A lot of calculations have been done on this. It turns out that the time when coal will be exhausted with the assumption of *prosperity* (and not depression) in the Western world countries will be around 2030-2050; let us say, therefore, sometime around 65 years from now.

But, sad to say, even this estimate may be too optimistic—it doesn't take into account the fact that there may be difficulties in getting the coal out of the ground at a sufficient rate. Thus, the amount of coal we shall be able to use in the next few decades may not depend only upon need, and how much coal there is in the ground. We should have to build a *very* large number of *new* mines to get the extra coal we will need, in time to take over the main energy burden. Correspondingly, we should also have to train an immense number of new miners—about two million! The demands of these factors may be too great for us. This sort of difficulty in expansion in the few years left may determine the amount of coal we could use, limiting its potential as a reliable alternative to oil and natural gas.

More Pollution?

Another problem with coal is that it is a gritty and dirty substance which gives rise to a lot of pollution in the atmosphere, including dust, organic compounds, and also, because it contains sulphur, sulphur dioxide, a material which is odiferous and irritating to the eyes.

What Would Happen If We Stopped the Growth of Our Economies?

An energy shortage threatens us sooner than anticipated because of the dynamic growth of our economies. If expansion were not tradition, it would be easy to go on with the use of the present fossil fuel sources longer than the times (around the year 2000) we have stated above. In particular, coal could last more than 100 years in several countries, and time for the development of the renewable, clean, abundant, energy sources would be extended by several decades. However, there are a number of reasons why it is difficult to slow down expansion of an economy.

For one thing, in the Western world, we live in a political system in which the idea of "personal freedom" is regarded as so important that we accept virtually any of the penalities of this freedom rather than allowing it to become restricted. In terms of living standards, personal freedom in Western countries means that the degree of freedom is controlled by the amount of money earned above what is necessary for the basics, with few interferences and restrictions imposed by external political forces (i.e., the government of the country). For example, it would not be acceptable politically for a government in the Western world to tell people that they must not have more than, say, two children per pair. If a government did this, people would say it was interfering with the freedom of the citizen. Therefore, the population will go on growing in many countries (although it has reached a constant value among the disciplined and highly educated Japanese). If the population goes on growing, the energy needs per year will continue to rise, and our present fuels will exhaust uncomfortably soon.

Again, much of the promise of our type of consumer oriented have-it-now society is in the expectation of the raising of material living standards each year. Everybody wants to have a car or two, with a trade-in for a newer model every three years or so, as well as a new color television set, large numbers of automatic household appliances, and so on. And they want to have more of these things each year, or improved models. That is one reason why they can be stimulated to work hard—to get the money to buy the goods from which they seem to derive so much pleasure.* If we limited the amount of consumer goods among the parts of the population which do not yet have all they want, some of the dynamic, successful character of the Western economies, wedded to ever-increasing consumption and the consistent expectation of increased living standards, would be gone. For why should people work as eagerly as they do in Western countries if they are not going to get an increasing reward? Many people do not particularly enjoy their jobs, and even the work ethic cannot explain the high incidence of ulcers, heart disease, etc., which people are willing to put up with in order to get ahead. Without financial reward, and the increased social status it allows, the carrot would no longer be there—only the stick would remain, a less attractive life—a life, in fact, similar to that in socialist countries with planned economies, where extra work and enterprise are not rewarded preferentially, and where, therefore, there is a slower tempo of life.

Thus, we probably cannot simply stop our economies from growing without giving up some personal freedom and reducing the individualistic Western life style: we should rather do practically anything than that, although a limit to growth must eventually come (even energy intakes from renewable sources such as solar energy are limited, in terms of production *rates*). Figure 4.8 shows the future of our energy alternatives.

Is Conservation the Answer?

The answer all depends on what is meant by conservation. If we mean "guarding the natural environment," then it is a good idea, though not directly related to the questions being discussed in this chapter. However, if, by conservation, we mean a cutback in energy demand, the term is very important here. It has been proposed that, for example, we do not *really* need two-car families, etc.

We could indeed save some 10% of our energy demand by a rational discontinuation of the energy waste which abounds in our spend-and-throw-away society. But cutbacks greater than this, say, 25%, would significantly decrease living standards. Two-car families may not be necessary

---

*Indeed, it sometimes appears that it is the buying of the goods which provides the most entertainment—shopping itself has become almost as important as the goods purchased, due in part to emphasis on advertising, attractive packaging, pleasant shopping environments as in suburban malls, and the opportunity it provides for getting out and socializing.

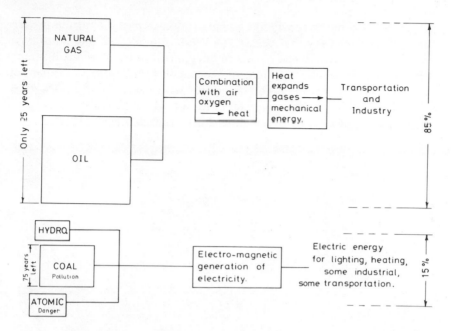

FIGURE 4.8. Oil is our major source of energy at present, followed by natural gas, then by coal, with hydroelectricity and atomic energy as minor sources at present. The diagram shows, very roughly, the years left for the main sources. Note, however, that we do not at present have the means for converting coal to suitable fuels on a massive scale.

(are TV's or dishwashers necessary?) but if you have one car and two adults in one family, each with different jobs in different parts of a town, it's decidedly awkward. "Decidedly awkward" means a lower living standard.

At about a 50% cutback in energy per person, life would be entirely different, crippled compared to its pace and scope before the cutback, and at a 75% cutback, one would approach the more primitive communities of the 19th century (90% cutback would mean basic village life of 1000 years ago).

All this has to take into account population growth. To stay constant in living standard a country must increase its energy supply by the percentage increase in population. Negative population growth (*falling* population) is, of course, an important conservation matter.

Shall We Run Out of Fossil Fuels in Our Time?

To finally answer this question, it must be clear that by the phrase "in our time" means within the next thirty years or so, because most of the people reading this book will live to see the decade past the turn of the century. We can then say that yes, in the Western world, oil and natural gas will exhaust around the end of this century. With respect to coal, it is more difficult to make a definite statement because of the difficulties of expansion and mine building and pollution associated with the use of this resource. If such difficulties can be overcome (and this would not be easy) the answer is no, coal will not run out in our time, but in our children's lifetime.

5

How Long Does It Take to Develop and Build Up a New Technology?

INTRODUCTION

The following two points summarize those discussed so far in this book:

1. The energy supply is as important to a technological people, an affluent people, as the air supply to biological man. Seriously decreasing the consumption per person would cause economic depression, then disorganization, and, if the decrease were sufficient, disaster. A significant decline in the standard of living (rise in unemployment, increase in inflation, decrease in real income), compared with that of 1973, was already visible in the second half of the 1970s in several Western world countries. One of its causes was the increase in the price of energy which was occurring because of the exhaustion phase entered by the world oil resources.

2. We are running out of the sources of energy which we now use, and these will be exhausted within two to three decades. After that, we could burn all our coal (despite pollution, increased mining difficulties, and costs, etc.) which would extend the lifetime of the affluent and expanding economies a few decades more.

When speaking of decades, it may not be clear whether this is truly a short time or not, in terms of an energy crisis. Certainly, it includes the lifetimes of most of us, but why are we stressing the urgency of the crisis? Is a decade really a short time, or is it actually more than enough for development of the energy alternatives we seek? It is easy to find out what counts as the yardstick of time in the present context of how long the exhaustible fuels will last. It is the time to build a new energy system of the *inexhaustible* resources.

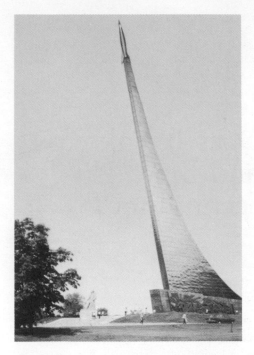

FIGURE 5.1. This is a statue in Moscow, U.S.S.R., which shows the ascent of the first vehicle which would orbit the earth. At the bottom stands the inscription, "First one must learn to dream."

STAGES IN DEVELOPING A NEW TECHNOLOGY

We can consider this generally at first, and then apply our discussion to energy technology. There are four stages:

The First Stage—Dreams

The first is the most exciting stage to a scientist or inventor, and without it, nothing can begin. "First one must learn to dream" are the words at the bottom of a famous statue in Moscow which shows the ascent (see Figure 5.1) of the first space vehicle. Unless one dreams and has ideas, without thinking of economics, public acceptance, etc., one cannot ascend to something new.

New ideas often enter the scientist's mind in the dream (daydream) stage, followed by a few years of discussion among scholars of the world, in universities and at international scientific meetings all over the world. Gradually, the *ideas* for a new advance become clear. Written up in scientific papers, the new idea (the potential base of a new technology) has then reached the level at which various approaches from applied technologies compete to deliver the *cheapest* way of producing something the public will wish to buy.

The Second Stage—Fundamental Research

Next occurs the experimental research level. Teams of scientists and technicians work together, often in laboratories at universities or in industrial companies, doing vital differentiating experiments to test ideas of the scientists who first created the hypotheses.

This is the stage of research paper writing. Scientists write papers describing what they have found, and the papers are communicated throughout the world through publication in scientific journals. Other scientists read and discuss them, the ideas spread and multiply to form new ones, and research is developed to test them further, as the work progresses. Of course, this stage is dependent on funding either from government agencies or private organizations, in addition to support provided at the universities or institutes themselves.

How Long Does the Research Stage Last? Research on a subject can go on for decades, or even a century. However, at the same time, a point will come, after a few years or so, when there is sufficient knowledge from the basic laboratory work for people to enter the third stage of putting the idea into useful practice.

The Third Stage—Developmental Engineering

This stage also involves research. However, the work no longer continues on such a small scale, on the laboratory bench, but in very big experimental setups in which the actual, useful result of the idea is built and experimented on. It is called the "development stage," or the "pilot plant" stage.

Here, for the first time, ideas which gave rise to the fundamental research approach reality. Micro- and mini-versions are built of what, one day, will become large plants serving the public by producing things they need or want. Modifications are made, each new version is tested and continually readjusted, and some ideas are scrapped. (These may have looked good on paper, but turned out to be too expensive or plagued by too many difficult problems to pursue.) This stage involves many complexities and can last for several decades.

The Fourth Stage—Commercialization

Commercialization means development of the technology for the public. Before this stage, less attention is paid to excessive costs, noise, pollution, durability, practicality of operation, economics of production, servicing, etc. During commercialization, the impractical, overly expensive, or dangerous flaws are removed. This stage includes mass production, packaging, and producing the instruction manuals and servicing organizations, etc.

In developing a satisfactory electric car, for example, the idea stage might involve the appropriate combination of hydrogen fuel cells as the energy source and a certain high-density energy battery for the power source.

The development work is connected with testing the power system to see how long it lasts, finding its reaction to driving cycles, developing the correct kind of light-weight motor and a system of magnetic braking, etc. This work is done in engineering research laboratories on simple prototype models of the car: 4 wheels, a body, the motor, and the power sources.

Commercialization would involve all the rest, and is worth doing only if the development stage is very promising. The car body is built, paying attention to external and internal details, design, and comfort. Refueling stations and servicing networks are developed if those already existing are not suitable, i.e., methanol or hydrogen stations, battery recharging areas, etc.

Finally, after the stages of ideas, fundamental research, developmental work, and commercialization, production can begin on a large scale, and the product sold to the public, so long as such sales (as the commercialization research will have shown) will profit the company.*

How Long Does the Commercialization Stage Last? There is no set answer to this, not even in terms of decades, because commercialization and the production of successive improvements continues as long as the technology has value, and as long as the producers can make a profit and thus have the incentive to operate. For example, government taxes, import–export regulations, or price controls, can slow down production of products.

The Cost of the Various Stages of Developing Technology

The single most important practical consideration in developing a new technology is money. How much will the research stage cost? It will be paid for by government funds, which come from the tax monies the government receives; so the more the government has to spend on research and development, the less net income remaining for the individual citizens.

Actually, the fundamental research stage is cheaper to pay for than the other stages in the development of a technology. Single, comparatively

---

*Profit is the catalyst of progress. Its prospect switches people on and makes them work harder. In countries where there is no profit motive, progress is slower in commercialization of a new technology.

small instruments and equipment items are used by a few scientists, and perhaps a few technicians. The cost of this work per year is only about double the salaries of the scientists concerned, the rest being used for equipment, items consumed, technician service, etc.

The development stage requires machinery and equipment of larger proportions. Here, the cost of the materials matters very much. We should expect that expenses at the development stage could be several times more than those of the research stage. The cost of the commercialization stage and, finally, that of the building of plants and technology around the country, will cost very much more than all of the previous stages. But, by the time we get to the commercialization and building, we have confidence that we are going to be successful, i.e., make a profit. In the cheaper, fundamental research stage, this is not the case. Hundreds, perhaps thousands, of research projects are started, but only a few lead to something economically acceptable. The development stage, too, often shows that devices concerned cannot be produced at acceptable price or safety levels, and the work may then be stopped.

Investors judge a project principally after the development stage has progressed somewhat. They are willing to invest capital in the idea if they think the chances of profit-making are high. The financial burden is thus shifted from the government to private investors. The public ends up paying in any case, through taxes as well as through purchasing of the marketed product.

Historical Guidelines Tell Us How Long Technologies Take to Develop

We can look at technologies already developed and find out how long they took to mature. Examples are given in Table 5.1, where we can see figures for the development of electricity and other forms of energy. Roughly a quarter of a century is needed for a product to be developed. It has taken two to three decades to bring inventions of the past to fruition and service. The time for building and the general spread of a technology is not included in this estimate.

It Is Likely to Take 25–50 Years (1–2 Generations) to Develop New Energy Sources

It seems likely that our new energy sources, and their time of development up to the commercial level, will take 25–50 years. Only recently, in the

TABLE 5.1. TIME DEVELOPMENT OF SOME ENERGY SOURCES

Source	Scientific feasibility established	Useful power	Economic power	Time
Electricity	Faraday (1831)	Sturgeon (1836)	Siemens (1856)	25 years
Steam engine	Newcomen (1712), Watt (1765)	Many developments	1785	20 years
Fission	1942	1955	1965	23 years
Fast breeder	1950	1960	1980-1900?	35+ years
Fusion	1975?	?	?	
Solar	1954	?	?	

mid-1970s, has intensive research on renewable energy alternatives for use on a massive scale begun. One of the tools useful in this research is the electron microscope, shown in Figure 5.2.

MONETARY ASPECTS OF BUILDING A NEW SYSTEM

Investment in fundamental research is risky, for this kind of research often results in failure from the investor's point of view. Because fundamental research is a basic activity for all communities (the main antidote to stagnation), it will be funded by the government. For commercialization (stage four), people (companies, investors, bankers) are willing to invest their money because there is a good chance that they will get richer by doing so. However, the commercialization stage of a great technological change, such as a change to solar energy, may require vast sums of money, equal to those required for building several cities. Getting investment capital of such magnitude may require a great national plan, and a fraction of the gross national product greater than the defense budget.

So, one problem will be the rate of investment during the commercialization stage of a new energy system. This will be one of the causes which will limit its rate of completion. Another cause of delay, apart from the limitation of investment capital, will be the number of qualified engineers and technicians available, as well as the availability of building materials.

A NEAR THING?

In Chapter 4, we estimated how long the present energy sources will last. You may recall that much depends upon the rate of growth (or decay) of the economies. It is this growth which is making us run out of resources more quickly than we had expected. If we cannot be ready with development of renewable energy alternatives in time, before our fossil fuels run out, then, on the basis of the present rate of expansion, we may have to accept a temporary decrease in our standard of living, with a longterm world depression. In this way, with the fall in living standards, we would use less energy per year, and hence our fuels—our old, polluting fuels—would last longer. We would then have more time to develop and build the necessary new system of clean, inexhaustible energy. But a decrease in economic growth will result in increasing unemployment.

We come now to the question of politics, democracy, and the needs and wants of the average citizen. People are more willing to be taxed for ends they see as being justified, particularly in a war, where fear or anger are aroused. But they won't have their personal life styles dictated by the government when there are no *immediate visible* energies.

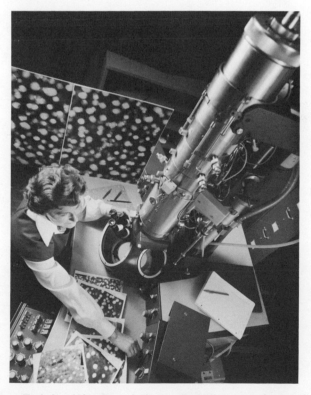

FIGURE 5.2. Anna Turkalo of the General Electric Company, researching with an electron microscope.

Zero energy growth will be as difficult to achieve as zero population growth, and the two are very much interrelated. Zero population growth would slow down the increase in energy need and prolong the supply of fossil fuels until we have done the development and building which will enable us to live on the renewable resources of solar and fusion energies. However, a large number of people may be willing to slow down their energy consumption, just as many have already decided on smaller families, if they see a true need. A distrust of both politicians and business corporations has created a reluctance in the public to sacrifice any part of their lives for a "maybe" crisis, a "maybe" energy alternative, and a "maybe" future.

Part II

Alternatives:
What Could Replace
Our Exhausting Fuels?

There is not much knowledge and information in the community at large concerning sources of abundant clean energy that could replace the present polluting and exhausting ones. One common view is that we simply need further exploration to find more oil. Such thoughts are true in the sense that a few new sources have yet to be found and exploited; however, it must be remembered that these sources have already been included in estimates of the date at which oil will be used up.

Others believe that there may be enough coal to last through at least twice their lifetimes. There is indeed enough coal to last two more generations, but we probably will not be able to get it out of the ground quickly enough to replace oil or natural gas (see Chapter 4). Further, the general use of coal to replace oil would bring much air pollution, because of the introduction of sulfur dioxide, a potent eye irritant, into the atmosphere.

WHAT ENERGY SOURCES SHOULD BE PROHIBITED FROM REPLACING OUR PRESENT, EXHAUSTING, POLLUTING SOURCES OF ENERGY?

It is worthwhile looking at this matter of coal as a future energy source a bit closer again, because many people believe we can rely on coal as the bridging energy source from oil and natural gas to clean, inexhaustible sources of energy, the technology of which we shall have to develop. What advocates of coal want to do is to combine coal with steam to produce methane, the main constituent of natural gas. If methane were obtained in this manner, we would call it "synthetic natural gas," a strange name because of its inherent contradiction of terms, but justified by the manner in which it's obtained.

There might be other possibilities with coal, too. The Germans have been forerunners in this respect. Gases which result from the reaction of steam with coal are passed over a special substance, called a catalyst, which accelerates the reaction, in this case working at high temperatures and pressures. Complex reactions then occur since the gas contains a great deal of larger, heavier hydrocarbons, such as octane and decane. These materials are amongst the constituents of gasoline (high in octane), and diesel oil (high in decane). So, we could get "synthetic" natural gas and "synthetic" oil, and hence gasoline, from coal. So far, so good.

But there are difficulties in developing a gigantic coal industry to supply our needs in place of natural gas and oil by 1990 or 2000, when such gigantic coal gasification would have to be in full swing. One problem lies in digging a large enough number of coal mines in time and persuading about 1% of the population to become miners! This would be a mighty task. We would have to open new mines every *day* in the United States until the year 2000 if we wanted to become completely independent of oil and able to rely basically on coal by that time.* This is a tall order, and probably an impossible one. Even if we could get away with building fewer mines by building much bigger ones, say 10 times bigger than the average ones now present in the United States, a new giant mine would have to be opened every second day until the year 2000—an impossible task.

A second difficulty is associated with the proposed future massive use of coal. Even if we managed to overcome the mine-building difficulty mentioned above, and contrived to open enough mines in time, or found some way of releasing energy from coal while still in the ground, the difficulty is the pollution which would occur if we used coal on the large scale necessary for total replacement of the other fossil fuels.

The major cause of the pollution would be the sulfur dioxide which is released into the atmosphere when coal is burned. This is because coal contains sulfur, and even if we invented a good method of removing sulfur from the granules of coal before it is burned, it would be difficult to store the vast amounts of sulfur we would have to extract. We would end up with a great many mounds of sulfur all over the world—another pollutant. To replace our exhausting, polluting oil with the even more polluting coal,

---

*Like many other statements about the future, this assumes that the growth of energy need will continue along the path it has pursued for many decades. It assumes that all oil and natural gas will be replaced by coal by the year 2000, with mines the same size as the present ones.

which would also exhaust, in a few decades, is certainly not an attractive alternative.

How Far Must We Look Ahead?

The matter of speed of development is crucial in obtaining new energy sources. We have discussed this in the earlier part of this book. *The critical time is the next 25 years.* That is a short time in which to make such a large-scale change.* In the following sections, we outline the possible paths to abundant clean energy, and in the rest of the chapters in this part of the book we discuss each of the alternatives in detail.

The Energy Sources We Should Develop to Replace Oil and Natural Gas Must Be Both Inexhaustible and Clean

Atomic Energy Sources

These are sources which have been mentioned frequently by policy makers in the recent past. For some 25 years, man has been relying on the prospect that atomic energy will take over the world's energy burden after the fossil fuels are exhausted. The way atomic energy works and the advantages and disadvantages of both fission and fusion reactions will be discussed in Chapters 6 through 8.

Direct Solar Sources

Solar sources are inherently abundant, and shall be discussed in detail in Chapters 9 and 10. There is much solar energy arriving on this planet: it is a matter of collecting it and converting it into a suitable medium of energy. If that can be done at an *acceptable* cost, we could live on this source for all time, and without pollution, as long as we can stop the population growth beyond a supportable limit.

Indirect Solar Sources

We have already discussed the fact that the fossil fuels are products of direct solar energy which have stored this energy through the process of

---

*One person's short time can be far beyond another person's horizon, and *that* is a considerable difficulty (see Chapter 18).

photosynthesis. This is a lengthy process, and is no longer able to provide us with energy for the future since our supplies are running out and are not being replenished. However, there are other sources of energy which *are* renewable, and which are also stored solar energy. Examples are wind energy and ocean thermal energy, which are discussed in Chapters 13 and 10 respectively.

Another type of energy which depends on the sun is hydroelectric energy. Due to gravitational pull of the earth, water will flow down a river and over a cliff, forming a waterfall. The energy from the fall comes from the stored potential energy in the water due to its height, and is given off as kinetic energy as the water falls to a lower level of both height and energy. The kinetic energy can be used to turn rotors which drive electric generators and produce electricity (see Figure II.1).

The role of the sun in all this is that it causes the evaporation of seawater, followed by rain, which replenishes the rivers and allows the waterfall to go on continuously (until destroyed by erosion, if allowed).

Other sources of energy which have all depended upon the sun at some

Figure II.1. The Tumet 3 Power Station of the Snowy River Hydroelectric Authority, New South Wales, Australia.

point include geothermal heat (heat from beneath the surface of the earth, or from hot springs), tides (also involving gravitational pull of the moon on ocean surfaces), and a variety of others, which will be discussed in Chapters 13 and 14.

6

The Dream of Cheap, Clean Atomic Energy

The Beginning of the Dream

Fundamental research for atomic technology began in the 1920s (see Figure 6.1), some six decades ago. In 1939, Professor Hahn and one of his students, Luise Meitner, found that very large amounts of energy in the form of heat were released when atomic nuclei were split.

Thus we began the fundamental research stage in atomic energy development. Then came the developmental stage, which was carried out during World War II under "pressure-cooker" conditions. The American government had been warned by Albert Einstein, after discussion with Niels Bohr, who was carrying out fundamental research at the University of Copenhagen in Denmark, that the discovery made by Professor Hahn and Luise Meitner in Germany could give rise to a city-destroying weapon, the atomic bomb. Profesor Bohr was removed from Denmark, which was occupied by German troops, in a British submarine, and flown to the United States. The American government put super-fast conditions of development into the practical realization of an atomic bomb, grouping together a large number of the best scientists and engineers whose entire time was spent working on the thinking about the project.

In atomic bombs, energy is produced in an uncontrolled way, causing very destructive explosions. However, the U.S. also stressed the aim of developing *controlled* fission reactors for the development of peaceful atomic energy. The enormous power of the atomic weapon, which could annihilate an entire city and its population using a single bomb (as demonstrated a few years later in the tragedy of Hiroshima) made the concept of the fantastically abundant energy of nuclear reactions vividly real to the public. The great difference in magnitude between the power of an atomic bomb and that of a

FIGURE 6.1. Marie Curie was the first person to extract radium, which was the first identified radioactive substance. She was awarded the Nobel prize for Physics in 1903 and for Chemistry in 1911.

bomb based on a chemical reaction stimulated the people's imaginations. If we could burn atomic fuel steadily and slowly, we could obtain much larger quantities of energy per amount of fuel than obtainable from chemical fuels such as coal, oil, and gas.

A nonexploding device for developing heat from atomic fission (splitting) was realized for the first time at the University of Chicago in 1942. An Italian physicist, Enrico Fermi, led a team which produced heat from controlled, slow splitting of uranium atoms to its fission products.

It is essential to make clear the fact that the time span for the development of the fission reactor—1939–1942 for the fundamental research to the pilot plant stage—occurred *much* faster than the decades which usually pass before ideas materialize into pilot plants. This was due to the war-time "pressure-cooker" conditions under which the development took palce, and which seldom occur in peace time. After this brilliantly rapid beginning, under wartime dictates, progress slowed down considerably for atomic developments.

THE DREAM CONTINUES

In the period after World War II and into the 1950s, people had a special attitude towards peaceful atomic energy. It was the dream of the century, and perhaps one of the greatest dreams mankind could have. It was thought that by the mid-1960s, atomic energy plants would be in worldwide operation, providing very large quantities of cheap power. The utopian concept of a world run by machines, allowing men and women to devote their energies to pursuits of their own choosing, seemed within a generation or two of being realized. There was no premonition of an energy exhaustion disaster facing mankind. The only fear regarding atomic power was of atomic warfare, a very real threat, but separate from the subject of peaceful atomic energy, since the weapon aspects would continue regardless of peacetime application research.

A Quantitative Comparison

The excitement generated by the discovery of atomic energy sources is easily understandable. To provide a more quantitative idea of the difference between the amounts of energy produced by atomic and chemical reactions, consider the following examples. The chemical reaction for the formation of water from hydrogen and oxygen molecules is accompanied by the production of relatively small amounts of energy. Specifically, the formation of 18 grams of water produces 68 kilocalories of energy. All chemical reactions produce roughly the same amount of energy. Correspondingly, a typical atomic reaction involving a uranium atom's nucleus and a neutron to form two new, smaller nuclei and more free neutrons, produces *30 million* kilocalories of energy per 235 grams of uranium used. Thus, we see that an atomic reaction produces about one million times more energy than a chemical reaction. Or, in terms of fuel consumption, a *million tons* of coal or oil could be replaced by *one ton* of atomic fuel.

The Dream Fades?

Why, after the optimistic expectations of the 1950s, are the hopes and dreams of an atomic era not realized? Clearly, the situation in the 1970s and 1980s regarding large-scale production of atomic energy is not yet decided. Many atomic fission reactors are being built in various parts of the world. However, the faith in the fervor for the atom, and an atomic era, is declining. Whereas in the 1950s, people felt secure about abundant energy in the future, we now have fears for our energy future. There are serious apprehensions that, within a generation, we might not have enough energy to sustain the standard of living of our ever-expanding civilization.

In Chapters 7 and 8 and in the remainder of this chapter, we will try to explain this change in outlook, i.e., the reversal in attitude toward atomic power, and our energy future in general. We will describe the type of atomic reactors being developed and the way they work, and discuss the knowledge that has been acquired during the past two decades about the effects of atomic power production on society, and on our planet.

Types of Atomic Reactors

Three types of atomic reactors have been suggested as future power sources. Commercially available reactors use fission (splitting) of atomic nuclei to produce energy, and are common in the United States and other countries. The shining metal domes of fission reactors are now a familiar part of the landscape.* See Figure 6.2.

Another kind of reactor, called a *breeder reactor*, is primarily in the developmental stage, although a few small ones have been built, particularly in France and the U.S.S.R. Breeder reactors differ from ordinary fission reactors in that they make their own fuel (they do, however, need a prefuel material). They breed their own fuel as they operate by bombarding nonradioactive materials with neutrons, thus producing a fissionable product. Their advantage is that they would not depend upon the availability of radioactive fuel, being able to breed fuel from the more abundant nonradioactive elements.

The third type of atomic reactor, the fusion reactor, has been discussed but not yet constructed, since the theory behind it is still in the fundamental research stage. It differs from fission and fission-breeder reactors in that it produces energy by fusion, or "joining together," of nuclei, rather than from fission, or splitting. These reactors will be discussed in Chapter 8.

The Fission Reactor

Uranium, the fuel for fission reactors, is one of the heavier atoms of those known to man. It is heavy because it contains a large number of fundamental particles, protons and neutrons, in its nucleus. This heaviness is the cause for the unstable nature of the uranium nucleus, and for the nuclei of all of the heavier elements. Protons have a charge associated with them (arbitrarily called "positive"), and, since like charges repel each other, the protons in the uranium nucleus repel each other, and since there are so many of them in the large, heavy nucleus, an instability arises. The other major particles in the nucleus, the neutrons, are neutral, i.e., they have no charge associated with them, and act as a "glue," holding the nucleus together by separating the protons from each other.

If the nucleus gets large enough, say by addition of an extra neutron, the binding process no longer works. The nucleus becomes unstable and

---

*The dome is there to prevent escape of dangerous radioactive material in case of accident in the reactor. Much more about this later, in Chapter 7.

FIGURE 6.2. A nuclear reactor in the United Kingdom.

tends to split up slowly to form smaller "daughter" nuclei, usually releasing some free particles, such as neutrons. This is the process of radioactivity called fission.

Uranium—The Active and the Stable Forms

Uranium atoms exist in two forms. Uranium-235 nuclei contain 92 protons and 143 neutrons, and uranium-238 nuclei contain 92 protons and 146 neutrons, and are therefore heavier by the mass of three neutrons. Uranium ores, found mostly in Africa and Australia, contain uranium bound with oxygen, forming substances called uranium oxides. These oxides contain both the 235 and the 238 forms of uranium in constant proportions of 99.3% uranium-238 and 0.7% uranium-235. However, only uranium-235 will undergo the spontaneous fission reactions we've been discussing. Uranium-238 nuclei, which contain more neutrons to bind the protons together, are stable. Thus, less than 1% of the uranium in the world (the 0.7% uranium-235) can be used as fuel for atomic reactors.

It is possible to partially separate the two forms of uranium. A new mixture of the uranium-235 and uranium-238 is obtained which is enriched with uranium-235. Such a mixture will be "hot," i.e., a greater percentage of the material will be radioactive.

How Fission Reactors Work

The purpose of a fission reactor is to enable the enriched uranium fuel to undergo slow, *controlled* fission (not an atomic explosion), and extract the tremendous amounts of usable heat energy produced. Recall that during a typical nuclear fission reaction, neutrons are both consumed (to initiate the reaction), and released as a product of the reaction. In the fission of uranium-235 nuclei, every neutron consumed produces two new ones, which can then be reacted with two more uranium-235 nuclei, beginning a chain reaction as shown in Figure 6.3. This situation could quickly become explosive, so some mechanism is necessary to control the rate of the chain reaction.

Two conditions are necessary for an atomic reaction to become explosive. First, it must contain a high enough percentage of radioactive material (e.g., uranium-235) so that the chain reaction can be sustained. Second, the actual mass of uranium-235 must be above some value, called the "critical mass," or the released neutrons will escape from the fuel rather than colliding with other uranium-235 nuclei.

One way to control fission and keep it nonexplosive is to mix the enriched uranium ore with a diluting material, such as graphite. This keeps the uranium-235 atoms sufficiently apart so that their emitted neutrons encounter the innocuous graphite atoms more frequently than they encounter

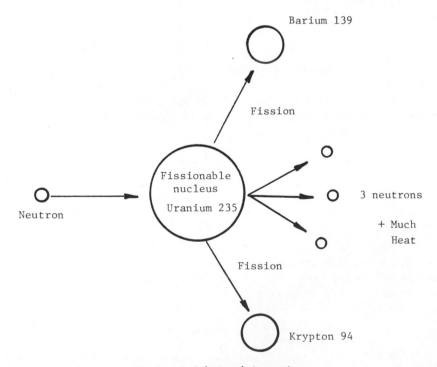

FIGURE 6.3. A fission chain reaction.

other uranium-235 atoms. This causes a slower rate of fission which makes the graphite and uranium fuel mixture glowing hot from the fission energy, but does not release the heat so quickly as to raise the mixture to explosive temperatures. It is allowed to glow for months without refueling. The heat energy is extracted by passing a liquid through the hottest region of the reactor in tubes. The liquid in the tubes is heated, and then led through water. The heat carried by the liquid is transferred to the water, which is vaporized to form steam. The steam is then used to drive electricity generators (Figure 6.4). One charge of uranium ore will last for about a year, and the weight or uranium needed in a reactor producing a million kilowatts is only about one thousand kilograms (about a ton) of uranium-235.

So, thus far, the fission scheme for atomic power production seems very reasonable and promising. However, a number of difficulties arise upon closer examination of the fission reactor idea.

The Difficulty with Nonbreeder Fission Reactors: Exhaustion

Even considering the vast amount of energy produced by fission of uranium-235, the uranium-235 atoms are used up, being converted to other atoms during the process. Since there is only a finite number of uranium-235 in the earth, it is an exhaustible, nonrenewable fuel source, just like the fossil fuels (coal, oil, etc.). The main difference is only that so much *less* of the atomic fuel is necessary, in weight, to produce the same amount of energy.

The question which logically follows is how much uranium actually exists on earth. A rough calculation can be done, assuming no new major deposits are discovered. The results show that if we converted to total dependence on fission reactors for energy, requiring about two thousand reactors, we would use up the uranium ore supply within *several decades.*

A possible solution would be to extract uranium from other sources than uranium ore. For example, it occurs in very small concentrations in certain mountain rocks called Conway granites. But the concentration levels are so low that we would need to excavate and process hundreds of thousands of tons of the rock to get a ton of uranium ore. The cost and energy expenditure in doing this would be prohibitively large, and at some future point we would still encounter exhaustion of uranium from these sources. Another possible source, the oceans, has the same problem of containing very low concentrations of uranium atoms, and we would face the same energy and money expenditures to get the uranium out. This problem of *diluteness* of an energy source will be encountered again in later chapters, in reference to other possible energy alternatives, including atomic fusion and solar energy.

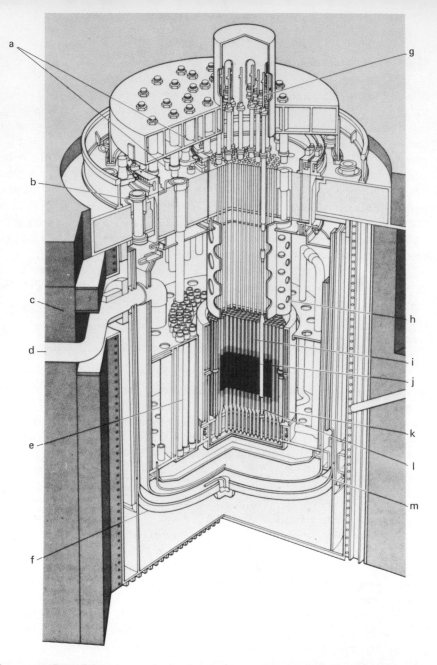

FIGURE 6.4. Part of the interior of a fast-breeder atomic reactor, depicted on the basis for a demonstration plant that would produce some 500 MWs of electricity. A full-scale commercial plant, scheduled for operation by 1984, would be of 1000-MW capacity. This design is of the loop type, meaning that the reactor proper, which is contained in a large tank of liquid solium, is separated from the primary heat exchangers and the associated pumps by loops of piping through which sodium coolant flows. *Key:* a, Fuel-handling ports; b, top shield plug; c, concrete support structure and shielding; d, sodium exit line; e, spent-fuel storage; f, insulated reactor-vessel jacket; g, control-rod drives; h, instrumentation support structure, i, axial blanket; j, active core; k, radial blanket; l, control rod; m, reactor vessel.(Acknowledgment to the United Kingdom Atomic Energy Authority.)

Thus, we are now faced with the decision of whether a few decades of fission-generated power will justify the expenditure of time, effort, and research, and use of other energy supplies which are running out (oil, coal, etc.), which are involved in the construction of the several thousand fission generators we would need to support ourselves on fission energy exclusively. It would seem that some energy alternative other than fission is not only desirable, but *essential* to continued existence at our present standard of living.

Will the Breeder Reactor Save the Situation?

Energy policy in the past has gone on the assumption that breeder reactors would extend the energy supply. The wisdom of this path is not at all certain anymore.

Instead of tossing away the idea of fission-generated energy, it is sensible to seek out other sources of fission fuels besides uranium-235, looking for one which occurs in much greater abundance. A device which is designed to achieve this is called a *breeder reactor*. It uses uranium-235 to convert uranium-238 into fissionable material which can be used as a fuel, thereby utilizing 99.3% of the uranium in uranium ores for energy production, instead of only 0.7%.

Activation of the uranium-238 occurs as follows. Fission of uranium-235 produces two neutrons. These neutrons encounter uranium-238 nuclei. The uranium-238 nuclei absorb the neutrons and become a new element, plutonium-239. Plutonium-239 nuclei are unstable, and will spontaneously undergo fission. Production of large amounts of energy accompanies this fission, which is a process very similar to the fission of the unstable uranium-235 nuclei.

The general breeder reactor scheme is shown in Figure 6.5. The fuel is now a mixture of both the radioactive uranium-235 and the stable uranium-238 forms of uranium. Uranium-235 is now used as a source of neutrons, rather than as a major energy producer as in a nonbreeder fission reactor. The emitted neutrons convert the stable form of uranium into a radioactive, fissionable energy source, plutonium-239. The discoverer of plutonium, Glenn Seaborg, is seen in Figure 6.6.

Ideally, if *all* of the nonfissioning uranium-238 could be converted into fissionable plutonium, we would gain a factor of 99.3/0.7, or 142 times our original 0.7% abundance of atomic fuel from uranium ore. Actually, the process is far less efficient, achieving only a little more than one plutonium atom for each uranium-235 atom used—in other words, the breeders are not really breeding. However, optimistically assuming 100% conversion of the uranium-238 into plutonium, i.e., 100 plutoniums produced per uranium-

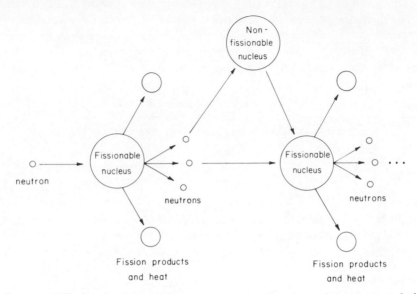

FIGURE 6.5. The breeder. Some of the neutrons produced can be used to create new fuel.

FIGURE 6.6. Glen T. Seaborg, Nobel Laureate, the discoverer (1941) of plutonium.

235 used, our utilization of uranium ore would be increased 100-fold, lasting for perhaps a few *hundred decades* instead of just a few decades. Or, assuming expanded future energy demands of about ten times our present demand, uranium ore could provide energy for about 300 years.

THE DRAWBACKS OF BREEDER REACTORS

We are again faced with some very important problems when breeder reactors are examined more closely. We already mentioned one—they do not seem to breed very well, if at all. However, even more pressing, and very frightening problems, deal with the effects of breeder reactors (as well as regular fission reactors) on us and our environment. First, disposal of radioactive wastes produced by the reactors has not been successfully dealt with, nor are any reassuring ideas in sight. Second, the possibility of accidents occuring has been shown to be a very real threat, due to the occurrence of a number of accidents in the past years. Finally, the effects of radiation either from wastes or from accidents or from hijackings have been shown to be very harmful and long-lasting, both to our environment and to our own health. These are all very serious problems, complex and *urgent*(because reactors are in operation, others are being built, and still others being planned), and they are discussed in the next chapter.

Fission Reactors—What Can Go Wrong

There are numerous problems associated with fission reactors, and they are *not* trivial. We are talking about accidents on the order of magnitude of nuclear attacks in wartime; cancer in the later lives of many of us; genetic defects in future generations of mankind; accumulation of waste products that could be lethal; and even the possibility of terrorist threats that could make airplane hijacking look like simple street muggings. We will discuss all of these problems in this chapter, and then explain why, in light of all these dangers, fission reactors are still being used, constructed, and planned.

FISSION REACTOR ACCIDENTS—A FEW CLOSE CALLS

What Can Go Wrong?

Fission reactors rely on water cooling systems, and it is this portion of the reactor that is crucial for maintaining safe, nonthreatening conditions. Were the atomic reactions in the core to proceed without proper cooling, an accident classified as a LOCA (loss of cooling accident) would occur. This is the most probable type of reactor accident.

If a LOCA occurred, the reactor core would fall to the bottom of its steel container, which is some 15 centimeters thick and could hold the core for a half-hour or so. The hot core would eventually melt through to the bottom of a second container, proceed to heat it, and again melt through in, perhaps, a day. The core, continuing to heat, would sink further underground and end up perhaps 100 m underground, migrating (at least for reactors in the

U.S.) in the general direction of China, and hence this sort of melt-down is referred to as the "China syndrome."

The mechanism of this melt-down will involve some blockage of the cooling water. The latter heats and converts to steam, and if the tubes containing it spring a leak, the resulting steam and water could descend and contact the uranium-containing rods. These are at a temperature around 900 °C, and the resulting steam would be explosively shot into the reactor container, taking with it a considerable amount of radioactive debris (from the result of contact with the rods). The steam is supposed to be contained by the dome of Fig. 7.1. Meanwhile, the internal structure of the reactor is increasing in temperature, its structure begins to fold and break loose—the China syndrome is on the way.

Now, the big question is: Will the dome hold? And even if it held, not blowing up with one big explosion, there is also the possibility of slow leakage. In 1973, the Swiss government considered purchasing atomic reactors from the U.S. First, however, they asked for experimental proof that the domes would hold in event of a LOCA, containing the dangerous radioactive materials produced. The proof could not be given, and still has not been demonstrated. An actual test would be expensive and dangerous (more dangerous than nuclear bomb testing, for it would have to be done on the surface of the earth), so all tests have been on paper, using simulated mathematical models. These results are of questionable validity, however,

FIGURE 7.1. An external view of a nuclear reactor. Note the dome, which might constrict the steam in a LOCA.

because they do not take into account imperfection factors due to human error, or engineering mistakes, shortcuts, or defective materials.

In other words, we have a situation in which we cannot do all the necessary testing, nor perhaps even anticipate what testing may be needed. Many unknowns are still involved in atomic reactors. We may be able to at least simulate and estimate answers to the questions we ask, but do we know *all* of the questions yet that *must* be asked?

If the dome does not hold in a LOCA in a populated area, thousands of people would die in a week from radiation poisoning. Tens or hundreds of thousands more would die in 20 or 30 years from radiation-induced cancer (see next section). And what sorts of genetic defects would occur in future generations?

One of the hardest-selling, proatomic reactor reports is the Rasmussen report, which states that the chance of a reactor accident is negligible, about the same chance as a major city on earth being hit by a meteor from space (about one in a million). The report uses simple reactor models, and neglects the imperfection factors mentioned above, particularly human error. The report says that multiple fractures would occur in boiling water reactors at a probability of 2 in every billion billion reactor years. In fact, over 15 reactor failures have occurred in a thousand reactor years.* A few of these accidents, including the most recent and perhaps most frightening, at Three Mile Island, are discussed in the following sections.

The Brown Ferry Mishap—Human Error in Action

In the 1970s, in Dakota, Alabama, a small accident occurred. A technician at an atomic power plant set out to measure the air flow from room to room. To test for direction of flow, he used a candle, observing the flame as he traveled about. Unfortunately, the flame came into contact with some urethane insulation, igniting the insulation around some cables. A fire spread to the control mechanism for the reactor-core water cooling system, also destroying back-up equipment which was supposed to take over cooling operations if the normal cooling mechanisms failed. Thus, a LOCA began.

As a result of the cooling mechanism destruction, the level of coolant water around the core fell from 200 inches above the core to 48 inches above the core, where it held steady due to an emergency arrangement. The cool-

---

*Thousand reactor years: 1000 reactors working for one year, or one reactor working for 1000 years.

ant systems were eventually repaired, and systems restored to normal. However, the LOCA focused attention on two very important examples of the human error factor:

1. The candle flame should not have been allowed to come into contact with the urethane insulation.
2. It was discovered that the fire extinguishers were not of a kind useful in the type of fire that started (the liquid was inappropriate), and the threads on the hoses and the extinguishers did not match each other.

In this case, emergency systems did eventually bring the situation under control, and the accident did not develop into catastrophe. Before discussing this point further, let's look at another, much more recent, nuclear power plant mishap.

Three Mile Island—"But No One Was Killed"

The Three Mile Island accident on the Susquehanna River near Harrisburg, Pennsylvania in 1979 apparently involved a reaction of overheated water with zirconium metal, causing the formation of a large bubble of hydrogen which interfered with the circulation of coolant water through the reactor. A degree of loss of containment within the dome led to a venting of some radioactive material into the surrounding air and some areas near the plant had to be evacuated.

One of the statements made by a defender of the plant personnel was "But no one was killed." However, the most likely fear from nuclear reactors is not violent explosion, but cancer induced by radiation from the plant. Such cancers take many years to develop; we do not yet know if anyone—or how many people—was fatally affected by the accident at Three Mile Island.

Biological Hazards of Nuclear Radiation—"Where Does It Hurt?"

Since the discovery of radioactive materials and the pioneering work and resultant radioactivity-induced deaths of scientists such as Marie Curie, the dangers of radiation to life have been recognized. Yet the crucial questions which remain unanswered are of primary relevance in discussion of fission reactors. It is not yet clear just what levels, if any, of radiation are safe, and what long-term effects can be expected from prolonged exposure to low-level doses of radiation.

What Increases in Background Radiation
Are Expected if We Switch to Atomic Power?

There is always a natural level of radiation which comes from materials in the earth and cosmic rays. Comparatively, the increases in background radiation due to atomic reactors has been very small. Even assuming use of 2000 reactors by the year 2010 or so, the increase in background level would only be 1–5% (aside from medicinal x-rays, etc.). For a long time, it was believed by most scientists that such small extra amounts of radiation could not possibly cause a health hazard because the levels still remained below those shown to be necessary to cause biological damage in laboratory experiments.

A remarkable change of viewpoint has occurred in these considerations the last few years. A Canadian physicist, Petkau, has shown that the damage done by radiations from radioactive substances depends not only on the total dose received by the body *but on the rate at which the body receives it.* The sense of Petkau's results is that if the dose concerned is given quickly (as in the laboratory experiments), it is much *less* lethal than if it is given slowly, i.e., received by the body over a longer time, as in a real situation of the exposure of the body to radiation in the air.

Thus, the laboratory experiments on the damage from radiations, upon which the building of the reactors has been justified, has given too optimistic a result. They had indicated (because the various doses were given quickly) that slight increases over background levels of radiation, such as one does get near a reactor, would do no damage to people. But it has now been realized that some damage to the body cells can indeed take place from the slight increase in radiation near reactors as our bodies are slowly exposed to these increased levels (see Figure 7.2).

Correspondingly, the ordinary background radiation (in earth's atmosphere) is now thought by some scientists to be the cause of various ordinary cancers, and perhaps even influences the processes of aging. Damage due to radiation, as described below, would promote diseases of the lungs, heart, and circulatory system, thus accelerating the problems associated with aging. In the presence of the slight increase in radiation which the reactors cause, the frequency of such cancers and other illnesses would increase.

CANCER-CAUSING CELL DAMAGE BY RADIOACTIVE SUBSTANCES

There are two mechanisms of cell damage by radiation (Fig. 7.3). The radioactive particles in the dangerous radiation are beta rays, which are streams of electrons.

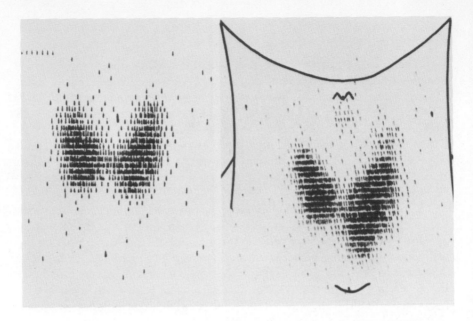

FIGURE 7.2. The detector scan on the left shows the distribution of ¹³¹I in a normal thyroid gland. The scan on the right shows the result obtained for a cancerous gland. (Courtesy of Oak Ridge National Laboratory).

The first mechanism is direct collision. The radiation particles strike the cells and burst them open. When cell growth, or renewal, continues after such damage, it may not be in the same direction as that of surrounding cells, and a cancer (which is the growth of cells in a direction different from that of the original normal body pattern) can occur.

The second mechanism is less direct, involving chemical reactions rather than physical, "brute force" destruction. When the radioactive particles enter the body, they collide with many water molecules (intracellular fluid is mostly water). The radioactive particles interact with water, producing fairly stable, modified, ionized forms of oxygen in solution. The water is decomposed, producing two bonded oxygen atoms with an overall negative charge. The oxygen ions then get adsorbed on the cell walls. Because the oxygen molecules are charged their electrostatic field affects the material of the cells.

The part of the cellular material of concern here is deoxyribonucleic acid (DNA), which contains the molecules which have within their structure what is called the genetic code, the information for the building of further cells. So, if it is damaged, further cells (still to grow) may grow differently, or "wildly," i.e., cancer may be produced. DNA is also the material contained in reproductive cells, ova and sperm cells, and damage to DNA in these cells would cause changes in future generations.

The Dose Rate Affects the Extent of Biological Damage

As shown in Fig. 7.3, if a given dose of radiation is administered quickly, there will be a great number of oxygen ions produced near a cell face at one time. Will these adsorb, i.e., attach themselves to the cell face and hence damage it? If there are a large number of oxygen ions at the same time (caused by rapid, massive doses), negatively charged oxygen ions will repel each other. Hence, they will not adsorb so well and will not do as much damage to the cell as when the same dose is absorbed slowly. In this latter case, less oxygen ions are produced per unit time, and hence each one

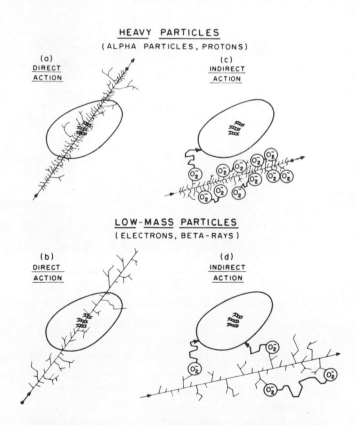

FIGURE 7.3. Mechanism of cell damage by radioactive particles.

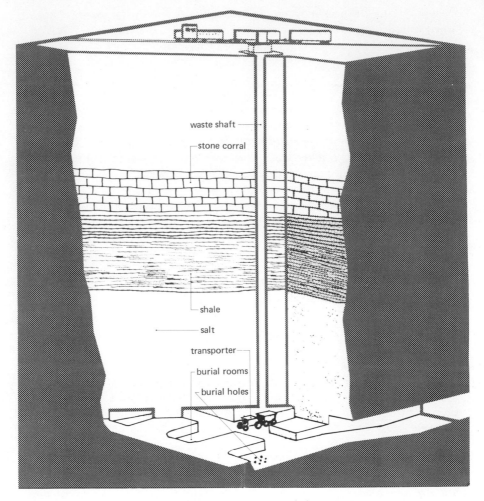

FIGURE 7.4. Disposal of waste in salt formation.

adsorbs more effectively on the cell wall (it is not repelled by another oxygen ion nearby) and damage is at a maximum.

Hence, it can be seen that the slow, long-term radiation exposure will promote the chemically induced cell damage, whereas quick, large doses are responsible for the physical rupturing of cell walls. Either mechanism can induce cancer and disease, and long-term low-level radiation is thus as dangerous as quicker, large-scale exposure. It simply has a delayed reaction, showing up perhaps 20 or 30 years later in life.

Dumping Radioactive Garbage in Whose Backyard?

One of the differences between atomic reactors and energy sources such as direct solar energy is that the atomic reactors leave debris that is harmful. Harm comes from the fact that it is still radioactive. Maybe it is less radioactive than it was when it went into the reactor but the so-called "spent fuel" is still far too dangerous to be allowed to be washed away, e.g., into rivers. Hence, one of the special difficulties of the source of energy known as "atomic energy," is "Where do we dump the wastes?"

There are several ideas for disposal. The first one is the one which the government is at the moment proposing. It is that the dangerous wastes should be kept or contained in glassylike materials and these should in turn be put into containers which are put deep into salt mines. These are conelike spaces under the earth which have cavities of very considerable size and have been undisturbed by earthquakes for many millions of years (see Fig. 7.4). It is thought that the slight degree of radiation which still emanates from the containers will certainly be dangerous.

This may be a good way of containing atomic wastes, but many objections have been made to it. For example, the containers get very hot; and what will the heat do to the stability of the salt mines? Will the salt gradually cause transfer from the sides of the domes to the metal containers and cause them to corrode and crack? There are other difficulties in the solution which must also be considered.

Some far more speculative and adventuresome suggestions have been made to deal with atomic wastes. Each has its proponents and its opponent. One, for example, would be to bury the material deep in the Antarctic ice. But will this someday melt and precipitate the material into the sea? Another concept is to put the material into space, perhaps to orbit the Sun and eventually crash into it. But how much would the orbiting of such materials cost, and further, what would be the chance that sometimes an atomic launch would go wrong and drop the material on earth, perhaps spreading over a few cities?

Last of all is one of the more imaginative ideas yet suggested. The earth's surface consists of various fundamental segments or sections, the surface areas of which are called tectonic plates: These areas gradually move over each other, and the concept is that if one were able to place containers containing the wastes into an area between them, they would gradually be "swallowed up by the earth," to be forgotten forever. This suggestion is perhaps a bit too far out at the moment for detailed discussion. Let us see.

8

Dreaming about the Future: Abundant Clean Energy from Atomic Fusion

FUSION?

For a hundred years or so man has obtained most of his energy by using chemical reactions, such as the reaction of fossil fuels like coal with oxygen of the air. The heat given out in such reactions is used to vaporize water, producing steam power to drive turbines and generators, and give us electricity. In the preceding chapters, we have discussed the prospects for a transition from chemically generated energy to energy derived from atomic fission. Very small masses of fuel would be needed to give the same energy as that given by large masses of coal in a normal chemical electricity plant.

There are difficulties in continuing to use the ordinary chemical way of getting energy, i.e., burning fossil fuels, because these fuels will soon be exhausted. But there are also difficulties inherent in the nuclear fission schemes. Exhaustion of uranium ore and the health hazards of extra radiation which enter the atmosphere near reactors, as well as the question of radioactive waste disposal, are problems which may prove unsolvable.

FUSION: ANOTHER ATOMIC ALTERNATIVE

There is another nuclear process which has a better prospect. It is called *fusion*. The meaning of "fusion" is "joining together" of two atomic nuclei to form a new, different nucleus. The nuclear fusion reaction we are most interested in for energy production is the fusion of two nuclei of a form of hydrogen called deuterium (which contains one proton and one neutron in its nucleus—the nucleus of common hydrogen contains only a proton) to give the element helium. A very large quantity of heat is produced during fusion reactions. In fact, the amount of heat given out is of the same magnitude as that which we obtain from fission reactions.

The Advantages of Fusion as an Energy Source

The advantages of atomic fusion as an energy source are considerable. The first advantage of fusion over fission is that the basic fuel needed for the fusion reaction, deuterium, is available in sea water. Deuterium itself does not exist in the form of free atoms and molecules, for it is combined with oxygen in the sea as water molecules. In the future, we shall have to reclaim it from the sea, which will involve decomposition of water molecules.

The Diluteness of Deuterium

There is much less deuterium than the ordinary form of hydrogen in our environment. About one-hundredth of a percent of hydrogen atoms found in water are deuterium. In terms of exhaustion, it does not matter that deuterium is such a small percentage of sea water because we have so much sea water available. One-hundredth of a percent of sea water is still a stupendously large amount, enough to supply our energy needs for millions of years.

However, the deuterium is very dilute. We will need to process huge quantities of sea water to obtain sufficient amounts of deuterium to fuel our fusion reactors. An important analogy to keep in mind is between the diluteness of deuterium in the sea, and the diluteness of solar energy reaching earth. We shall discuss in later chapters criticism of solar energy as a practical alternative because of its diluteness. Critics point out the vast area of solar collectors that would be needed, covering large portions of the earth's surface. However, it is less frequently noted that we would have to construct huge plants to process large portions of our oceans to supply our energy needs via fusion.

Is the Fusion Concept Utopian?

Despite the diluteness of deuterium, fusion appears to offer us unlimited energy for millions of years without the pollution and health hazards of fission and breeder reactors. Indeed, the cleanliness of the fusion reactions are a second major advantage of fusion over fission. Recall that one reason why the dream of plentiful atomic energy has been deemphasized is because the breeder mode of getting atomic energy produces plutonium, which, when accidentally released into air or water, is dangerous to life. But there is nothing inherently dangerous in the fusion process of forming helium from deuterium.* Helium itself is a benign substance. It is inert, i.e., does not have *any* chemical reactions, and therefore will not be active in biological systems.

This does not mean that there will be *no* danger in building the fusion machines which are at present only in the planning stage. However, there are no radioactive waste products to store as with the breeders. So here is an excellent prospect—the possibility of enormous amounts of energy with much less risk of radioactivity-induced cancer.

ATOMIC FISSION AND ATOMIC FUSION

Let us pursue the comparison between fission and fusion. Atomic fission looked wonderful to the people of the 1950s. So we must be careful not to pursue the fusion path enthusiastically without question and criticism, lest we find ourselves in the midst of another fiasco. In review, the advantages of fusion over fission as a process for getting energy on a massive scale are twofold:

1. With fusion, there is no question of running out of the fuel supply of deuterium. In fact, the time at which we should run out is the time our sun will have begun its final expansion at the end of its life. With atomic fission, on the other hand, we are limited by the amount of uranium deposits in the earth. There is not enough fuel (without the use of the controversial breeder) to rely on the fission process as a source of energy to replace oil and natural gas for even one century. There is only enough to last us for a generation or so early into the next century, after which we would run out of the uranium necessary to fuel the reactors. There is not much prospect, then, in trying to replace oil and coal with fission reactors.

2. Using atomic fusion instead of fission, there is a far lesser health hazard; for the reactant, deuterium, is harmless, and the product, helium, is not radioactive—it is an inert gas and utterly unreactive to living or-

---

*However, machinery used in the process of fusion may become dangerously radioactive itself and therefore be difficult to dispose of safely when it needs replacement. There is no 100% safe way to use atomic nuclear reactions. There will always be some source of unwanted radiation, in addition to the desired products.

ganisms. The dangers of fusion reside principally in dealing with the materials of the reactor rather than with product or reactant, a problem which will occur in any atomic energy scheme (see Table 8.1).

But Shall We Actually Be Able to Attain the Fusion Process?

When something is so advantageous, so utopian, as the fusion reaction, the prospects tend to bemuse man. The area becomes a wonderful El Dorado for seekers of research funding. Our enthusiasm sometimes dulls healthy, critical inquiry. The scientist-entrepreneur becomes like the hunter who talks a lot about the bears he has treed, but often, when asked to show the skins, changes the subject. So it should be admitted at this point that the fusion reactions, such as those shown in Figure 8.1, have not yet been attained *in a lasting way*.

Now, the *fission* processes described in Chapter 6 *have* been attained, and fission reactors are available and function quite well, unless an accident occurs, or they are not built properly or adequately maintained and supervised (see Chapter 7). In fact, about two hundred fission plants have been built throughout the world, and many more are in the planning stage.

The great hopes put into the fusion process are too idealistic. The situation looks less comfortable when one learns that the fusion of hydrogen nuclei to form helium has not yet been attained in a controlled way.

Of course, if the *controlled* fusion process really took place in practice, and the economic engineering problems had been solved, we would start

TABLE 8.1. Atomic Fission and Fusion: The Differences

Fission	Fusion
1. Basic process is *splitting* of atomic nucleus.	1. Basic process is *union* of atomic nuclei.
2. Products of process are radioactive.	2. Product is inert gas, helium.
3. Amount of actual ore in ground limited but very dilute concentration of uranium in many rocks.	3. Basic fuel (deuterium) is available in abundance.
4. Would need breeder process to use products of these rocks. Fission reaction commercial. Breeder process in pilot plant state, but as yet unsatisfactory, uneconomical, and dangerous.	4. Basic process not yet maintained for more than microseconds.

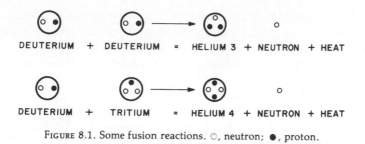

FIGURE 8.1. Some fusion reactions. ○, neutron; ●, proton.

building fusion rather than fission reactors. So, the *bad* news about the fusion process is that it has not yet been attained in a controlled way.*

Some scientists believe that fusion will be attained with a few more decades of heavily financed research. Indeed, if they did not believe this, we would not have been justified in talking about fusion as an energy source in the glowing terms we have been using.

DIFFICULTIES IN REALIZING A CONTROLLED ATOMIC FUSION PROCESS

Why is it difficult to attain a controlled fusion process? First of all, we have to attain a very, very high temperature to initiate the fusion of the nuclei. This temperature is around 10,000,000 °C. From the human viewpoint, 1000 °C is a high temperature, and by 2000 °C most materials have melted or decomposed. But temperatures 10,000 times higher than this are needed for fusion to occur. The reason is that when temperatures are very high, the reactant particles (in this case the hydrogen nuclei) fly around very fast, so that when the particles approach one another, their kinetic energy (energy of motion) is so great that the natural repulsion between the similarly charged nuclei is overcome. When the hydrogen nuclei bang into each other with such great force, enough kinetic energy is present to fuse them, forming a helium nucleus.

However, despite the high-temperature requirements, the situation is by no means hopeless. By means we are not going to discuss here, research scientists in the U.S.A., and the U.S.S.R. (they were first), and the U.K. (who were second) have been able to attain the necessary temperatures for fusion processes to occur, even though the process lasted for only a millionth of a second or so.

---

*It has been attained in an *uncontrolled* way, in the hydrogen bomb. But this is not of use to us as a *stable* source of heat energy which we can tap off and convert to electricity, although thought has been given to the possibility of setting off a series of small H-bombs on earth, perhaps in a valley in a remote place, and trying to collect the immense heat energy released.

It should not be thought, however, that because the first firing of the process has been achieved, that a stable, engineered fusion reactor is around the corner, perhaps only one or two decades away. The situation of fusion research is like climbing a mountain: it is increasingly difficult as the summit is approached. But, when (and if) we can cause fusion in a controlled way, we shall not have to continue heating the system externally. The heat given off by the fusion reaction will keep the mass of hydrogen and helium nuclei sufficiently heated so that it will remain reactive. This hot mixture of hydrogen and helium nuclei is referred to as a *plasma*.

One great difficulty of attaining fusion is raising the temperature of the hydrogen to the millions of degrees necessary to make it begin.

Plasmas

Even if we can get our plasma hot enough, there is a second difficulty in attaining a satisfactory fusion process. Before we explain the second difficulty, it should be pointed out that if the deuterium is to be at this high temperature it will not only be in the atomic form, i.e., existing as individual atoms rather than bonded to other atoms in a molecule, as at ordinary temperatures. It will also contain atoms in the *ionized* form, which means that the electrons normally around the nucleus in the atom will have been stripped off. The remaining atomic fragment will have a positive charge associated with it (recall from Chapter 6 that atomic nuclei consist of positively charged protons and noncharged neutrons, giving the nucleus an overall positive charge).

This state of matter, considered during discussions about fusion, is called the "plasma state," the state attained by an ionized gas. It is a mixture of electrons and atomic nuclei, which are no longer joined together to form atoms. Specifically, we are concerned with a mixture of electrons and *deuterons*, the nuclei formed when deuterium atoms have their outer electrons removed.

The Difficulty of Containing a Plasma in a Bottle Is That It Escapes

All chemical materials are easily melted or decomposed at temperatures approaching 4000 °C, in fact most substances melt or decompose by 2000 °C. A troubling question, therefore, comes into focus. How can we *hold* the plasma, the glowing ionized gas, at the ten million degrees necessary to achieve fusion reactions? For, if we cannot contain it in a fixed place, we shall not be able to control the fusion processes within it, much less draw heat or electrical energy from it.

There is no *material* which could contain such a glowing ionized gas, the plasma of deuterium. At first this would appear to be our much feared dead end on the road to obtaining massive amounts of energy by fusion. But there is an idea, suggested some years ago, for containing glowing plasma, which might one day overcome this second big difficulty (assuming we can first achieve temperatures of ten million degrees to start the fusion). This idea allows for containment of hot material at any temperature, so long as it consists of charged particles.

To explain this idea, we must first recall the principles of behavior of elementary charged particles (i.e., protons, electrons, and free atomic nuclei). When these charged particles are in a system moving at some speed and we apply a magnetic field (similar to the magnetic field of the earth's North and South Poles which attract compass needles) to the mixture of charges, there is a force exerted on the moving particles. The size of this force on a given particle in this mixture depends on the strength of the field we are applying, multiplied by the charge on the particle, and then multiplied by the speed at which the particle is traveling.

How does this solve the problem of containing a plasma? If we could have a magnetic field in the shape of a bottle or some container (and this is certainly possible), the field would grip the particles in the ionized gas, or plasma, because a magnetic field exerts a force on a moving charge and keeps it in its grip as the walls of an ordinary bottle keep a gas in an enclosed space (see Figure 8.2).

This "magnetic bottle" could be a solution to the major difficulty of how to handle and contain a plasma at 10,000,000 °C. Magnetic bottle containment is the central practical idea of fusion engineering, and one of the main supports for the fusion physicists in their hope of getting practical energy-producing machines from the fusion of hydrogen (deuterium) to helium. The plan is to exert a magnetic force on these charged particles and make the magnetic force exerting the field into the shape of a bottle or tube, which would contain the hot plasma, just as though the plasma were in a material bottle. A magnetic field is not a substance composed of matter (atoms and molecules), so it will not be destroyed by high temperatures.*

The likelihood and cost of engineering fusion reactors cannot be estimated until the plasma can be kept in a controlled space.

---

*One problem, however, with magnetic bottles, is containing the stream of neutrons which will also be produced in a fusion reaction. These neutrons will have a very high energy and hence will be destructive to any material with which they come into contact. To date, no one has been able to investigate the material effects of such streams of neutrons. From a theoretical viewpoint, the prospects here are poor. See Figures 8.3 and 8.4.

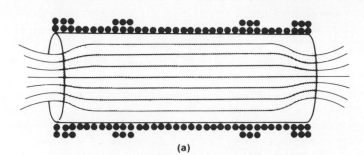

(a)

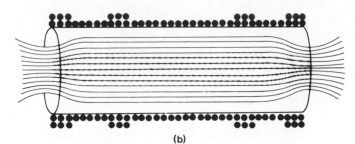

(b)

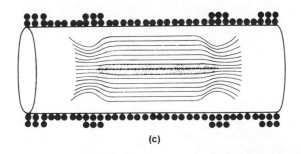

(c)

FIGURE 8.2. Adiabatic compression of the plasma. (a) Plasma is injected into the chamber while the field strength is weak. (b) The magnetic field strength is then increased, compressing the plasma toward the center and raising its temperature. (c) The magnetic mirrors may also be moved axially inward to provide additional compression and further increase in temperature. ●, energized coils.

FIGURE 8.3. Interior of a d.c. multipole, an early device development by the General Atomic Company to confine plasma in a controlled thermonuclear fusion reaction.

ANOTHER POSSIBLE METHOD OF GETTING ENERGY FROM FUSION

We have seen that the nuclear reaction in which nuclei of hydrogen isotope deuterium fuse to form nuclei of helium promises a great release of energy, with few health hazards from radioactive wastes. The high temperatures necessary to initiate fusion have been reached, although only with great difficulty. The main problem in deriving useful heat or electricity out of the reaction is that the plasma cannot as yet be maintained in one place, and hence the particles do not remain together at the high temperatures momentarily induced.

A totally different approach to fusion has therefore been devised, which completely avoids both the temperature and the containment problems. To explain the idea, we shall first discuss an analogous situation involving our sun. As explained further in Chapter 9, stars such as the sun are masses of plasma, undergoing the fusion of hydrogen atoms (of the ordinary form) to create helium. How did these stars begin these reactions? Gravitational attraction pulled bits of material in the universe together,

FIGURE 8.4. Doublet II, a device developed by the General Atomic Company, San Diego, California, in the effort to produce a leakproof "magnetic bottle" for controlled thermonuclear fusion reactions. The Atomic Energy Commission is supporting a $26 million program at the General Atomic Company to design and build Doublet III, which will be three times larger and more powerful.

forming a sphere which would eventually coalesce into a star. The attractive pull of gravity is proportional to the mass involved, so the attraction increased with the growth of the sphere. When the mass of the star got very big, the gravitational pull on each part of the sphere was so large that electrons were stripped off the atoms in the star. Thus, the frictional pull of atom against atom caused formation of a plasma, and the fusion reaction began. In our case, the result was our sun, our light, and the possibilities of solar energy (Chapter 10).

Now, this gives rise to an idea for useful fusion. Light contains a certain quantity of energy, which depends on its wavelength. Suppose we were able to produce very concentrated beams of light, providing enormous amounts of energy in each beam. This concentrated light beam represents a way of delivering a lot of energy very quickly to anything in its path.

Consider placing a drop of liqid hydrogen in the light beam's path. Instead of using ordinary hydrogen or deuterium, we will use a third form called *Tritium* (2 neutrons and 1 proton in the nucleus). It is advantageous to use this heavier form because it is able to absorb more energy than the lighter forms of hydrogen.

Now, we will deliver light to the tritium drop by means of a laser. This is a device which produces very concentrated beams of light, in thin pencil-like form. We allow a drop of tritium to fall, and at the same time fire the laser beam at it. When the beam hits the drop, if the drop is small enough and the beam intense enough, there will be a very strong delivery of energy from the light beam to the drop in a very short time, giving the drop an almighty whack!

What should happen? Well, if the whack is great enough, the tritium molecules in the drop will be broken down into atoms, and the atoms to nuclei, and the nuclei will be pushed very hard against each other. This is indeed the way our star, the sun, was made to enter the plasmic state and begin its fusion reactions.

So, you can see that we might be able to induce fusion by applying great *force*, rather than great *heat* (as in the first approach). As to the method of getting energy out, there would be much heat released when the particles in the drop fused to form helium, and there would be many drops, and many laser flashes. One can see that much heat could be produced.

The scheme for converting the heat energy into useful power is shown in Figure 8.5. The diagram is complex, but essentially what occurs is that the fusion-generated heat gets carried away in sodium metal, and finally the tubes carrying the heated sodium pass through water and turn it to steam. The steam can work an electric generator, and so on.

What Has Been the Progress of the Laser Method for Fusion?

This method is much newer than the magnetic bottle method, and a great deal of work is currently underway to test the idea for practical use. Generally, the fastest rate of progress on an idea is made when it is new (i.e., within 10 years or so). So, the remarks we make here are likely to be

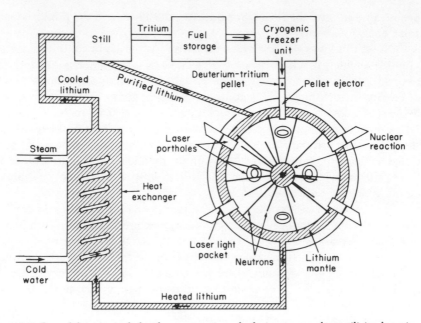

FIGURE 8.5. One of the proposals for the construction of a fusion power plant utilizing laser-induced fusion.

out of date by the time you read them. But there are two main difficulties found so far. First, it is not so easy to design a machine such that when the laser flashes a shaft of light (i.e., a beam of energy) intended to swipe the falling tritium drop, it actually hits it. There could be many misses.

Another difficulty is one which always has to be investigated for each new method of energy production: does the method produce a net amount of energy, or is the energy needed to make the method work greater than the energy produced? There is a danger of this being so in the laser method for fusion. Much energy is wasted working the laser, and it has not yet been proved that *net* energy is produced.

However, since the age of this method is very young, less than ten years, it is far too early to predict its eventual effectiveness. At any rate, it gets rid of the problem of escaping plasma associated with the magnetic bottle method.

So far, it has been explained that atoms of hydrogen in its various forms could fuse to form helium, and if this occurred over a significant period of time, enormous quantities of heat could be produced. In the laser fusion method, the ideas (and they are only *plans* as yet) for harnessing the energy have been explained. But by far, the most research has gone into the high temperature—magnetic bottle approach to fusion. So it is important to explain how energy would be extracted in a practical form, supposing the magnetic bottle could be made to hold the plasma stably.

One idea which could be the basis for massive electricity production is shown in Figure 8.6. The plasma from the bottle (a stream of deuterium ions and electrons) is expelled, and passes through two electrodes. At the first electrode, electrons are collected. They flow through an external circuit and here provide their energy to a practical, useful circuit (for use in town, factory, etc.). Finally, the electrons flow back to the second electrode of Figure 8.4 and neutralize the deuterium ions to re-form atoms. The atoms, when cooled, are again deuterium fuel, ready for the next pass into the plasma.

Is the containment of the plasma in the fusion of two deuterium nuclei feasible on a large scale? Although we have been able to attain high temperatures for fusion for a few millionths of a second only, we have experimented a great deal with trying to contain the plasma by the magnetic bottle method. The experience so far has not been good. At first it works, with the magnetic lines of force gripping the charged particles in the ionized plasma and pushing them into a shape like a bottle. Then many events begin to occur. The plasma becomes unstable and wobbles. It gets out of shape and tends to squeeze out of the magnetic bottle.

It is probable that with enough research these problems can be overcome. But at what cost in tax money? And in how many decades? With how many thousands of scientists, not yet trained? Processes which do not break

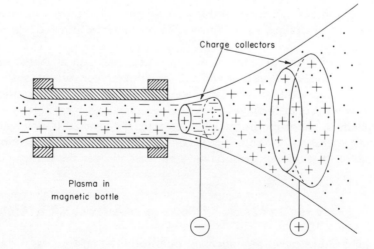

FIGURE 8.6. Direct electricity generation from fusion.

any fundamental laws of nature can usually be made to work with sufficient science and engineering research, i.e., money. Getting to the moon was thought to be ridiculous only a few decades ago. We did it. But we have not commercialized it yet, and that is likely to take a few decades more.

Time

Here we come to the heart of the problem. Fusion energy could offer a utopia in the future, just as we thought fission might, but the question is: How much research money would be needed? How many scientists and engineers would have to be trained for the area? How many decades would the development work take?

These are knotty problems. In particular, would we be able to get fusion going, engineered, and built, in the various countries, before the oil starts to run down and becomes prohibitively expensive in some 25 years? Or, even before the coal runs out, in some 75 years?

Trying to answer this type of question is similar to trying to predict the future career of a baby that has not yet been born. We cannot answer it with enough accuracy to be useful, since we do not have the knowledge to base an answer on. The process is not yet properly born, because the fusion reaction has not been made to function continuously, even in the laboratory. It is still in the fundamental research stage, on the drawing board so to speak. We are only just beginning to contemplate the pilot plant stage, and are only about half-way through the research necessary to begin such construction.

Who can say when the stable containment of the prolonged fusion reaction will be attained and engineered? Will it ever be attained? Probably. But when?

Not only is the attainment of controlled fusion in a vague, distant future, but, once developed, it will still take time to convert our systems to its use. We attained fission in 1942; we have only about 1% of our energy from fission thirty or forty years later. Thus, even if we were able to sustain controlled, stable fusion in the laboratory by the year 2000, it might take another generation or two to get it going in large plants *at an economical price*. After that, systems still have to be built all over the world.

Is Fusion the Best Energy-Producing Prospect of Them All?

Our heading poses an important question. The answer has to be *yes*. Fusion appears to be the most promising idea we have in the entire energy field.

Thus, the hydrogen (deuterium, tritium) fusion process may be one of the great concepts of mankind. In the long run, it could become our primary energy source. Then one could have small energy producing stations of enormous power built under cities. One such station, no bigger than present electricity generating plants, could supply all the energy for the entire city, not just its household electricity. One would require a link to the sea, or to other sources of water, and there would have to be an electrochemical process which would extract deuterium from water. After the fusion reactor had made electricity, electrolysis of water to hydrogen on a large scale would give hydrogen fuel, e.g., for transportation and to run industry (see Chapter 10). Thus, at the center of each community, there would be a large energy source (of small size), with no danger of health hazard, explosions, hijacking, or pollution. This would put us a giant step nearer the materialist utopia—attainment of the expectations we had in the 1950s with respect to atomic energy in the form of fission.

Certainly, the fusion of hydrogen to helium has advantages over solar energy because the space requirements for energy production are much smaller. There would be no problem of having to use large slabs of land for collection and no problem of having to send energy thousands of kilometers or more from sunny climates to regions of the industrial parks.

The atomic fusion process of hydrogen to helium may well be the great energy-producing method of some future era. But oil and natural gas will become relatively inaccessible in a mere two or three decades, due to their limited supply. The new sources of energy we should be developing must be the ones where success in the development work is expected on a time scale interesting to us.

Fusion Compared with Other Abundant Clean Energy Sources

We cannot wait 50 or 100 years for a new energy alternative. Fusion is a Rolls-Royce source. We could buy a Chevrolet sooner, and still not have to go back to bicycles.

For example, in Chapter 13 we will be discussing wind-generated energy. We could be putting up large rotors on the sea over large areas of the ocean in positions where the yearly average wind velocity is high. Problems would arise, because vast areas of the sea would be less accessible for shipping, and the danger of storms to the rotor devices would have to be overcome. But, on the other hand, the engineering does not need to go beyond principles which have been known for the past century, and no new basic research is required. We *know* and *understand* the problems. We also *know* we can build successful aerogenerators, on the scale we require, given per-

TABLE 8.2. THE CONSIDERABLE TASKS WHICH REMAIN IN
ATTAINING THE FUSION REACTION

1. Find easier method to attain temperatures of 10,000,000 °C.
2. Find out how to keep the plasma (if it can be formed) in a controlled space and
   hence keep tne fusion reaction occurring.
3. Stop the γ radiation that is given off, which is damaging the plant.
4. Engineer actual reactors which will compete in costs of final fuel with ocean
   thermal gradient and power tower solar-hydrogen producers.

haps a decade of intensive development work. This means building a pilot
windmill, pilot plants, and trial full-scale plants and testing them. But the
task of building large-scale wind collecting stations is very small compared
with the immensity of the task of building practical, economic, controlled
fusion plants on a large scale since we have not even made fusion work
stably in laboratories. Table 8.2 lists the tasks which remain before fusion
can be implemented on a large-scale basis.

So, perhaps we should end this chapter about the utopian dream of
energy via fusion by saying this: controlled fusion of hydrogen to produce
helium is the best idea for provision of energy for mankind on a long-term
basis. But it is still in the laboratory and has been faltering there for a while.
The fusion reaction has not yet been sustained after a quarter of a century
trying. Will it take half a century more to achieve that research goal? We do
not have the time to wait for this to be done. We need a new energy system
built and running in 25–30 years. Therefore, we must emphasize the development of ideas we know can succeed: solar, wind, gravitational, geothermal. Even these energy alternatives will require a great deal of time and
effort, but the amounts are predictable, and the results are likely to be successful.

The Most Available Energy Source: The Sun

Energy from the Sun

Our sun presents us with another form of energy, in addition to those originating on earth—namely, sunlight. Daylight is radiation which reaches us from the sun. Moonlight consists of light which travels from the sun to the moon and is reflected back to us on earth (Figure 9.1).

Our existence depends on the energy of the sun. In ordinary life, nearly all of our warmth and light comes from it, either directly or in a stored form from the earth's coal and oil, which are stores of material formed by the sunshine (and the photosynthesis reaction it supports, see Chapter 1) of earlier epochs. Before asking whether we can harness energy from the sun, converting its energy to a storable substance, we must ask for how long the sun will keep on burning.

The Sun's Expected Life

As we look at the sky at night we see a vast number of light sources. Nearly all of these are stars (the rest are planets of our own solar system, which reflect light to us from our sun just as the moon does). Most of the stars we see are suns which exist in our own galaxy, the group of stars in which our sun is located.

On a clear night, billions of suns which make up our own galaxy fuse into a continuous, luminous, cloudlike image in the sky, referred to as the Milky Way. Our sun itself, an average star in this galaxy, is surrounded by nine planets, including earth, and thus forms a solar system.

The most accepted idea of how the universe began is called the "Big Bang" theory, and according to this theory, the Biblical phrase "In the be-

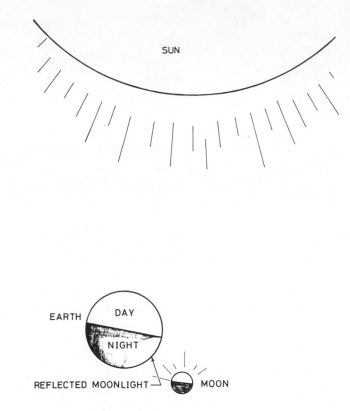

FIGURE 9.1. Sun lights the earth during the day. The moon sometimes lights part of the earth during the night. However, the moon's energy comes from the sun, so moonlight is just indirect sunlight.

ginning was the Word" could be replaced by the phrase "In the beginning was a super-gigantic sphere of neutrons." This sphere, according to one school of thought, was struck by a gigantic packet of light. This light caused the breakup of the sphere into a stupendously large number of fragments of glowing gas, mostly hydrogen. Because these fragments were great in mass, they could pull, by gravitational force, other atoms to them. This pulling caused friction. Eventually the gathered mass was large enough so that the forces acting were sufficiently large to cause the hydrogen among the atoms to fuse into helium (the same reaction we are trying to reproduce in the laboratory under controlled conditions, to form the basis of fusion energy production—Chapter 8).

The stars are, in fact, glowing masses of plasma, i.e., nuclei of atoms (mostly protons) and electrons, which radiate energy resulting from the fusion reaction.* It is the light emitted, i.e., the visible portion of the radiated energy, which enables us to see stars like our sun. Thus, from our sun we get both light and heat energy. (See Figure 9.2.) The heat energy which reaches this planet from the other stars is negligible.

We might call the glowing star mass a super-large fusion reactor, passing through space. It does not need a magnetic bottle (see Chapter 8) to confine it, because the huge mass is enough for the gravitational attraction (proportional to mass) between the particles within the plasma of a star to keep them all together in one massive fireball.

Suns keep on burning because of the energy emitted in the fusion reaction of hydrogen to form helium (Chapter 8). The difference between the controlled fusion reactions we hope to create on earth and fusion reactions which take place in the sun is the great mass of the latter.

The sun is not the calm sphere it appears to be. It is a super-giant mass of fire and intense heat, millions of degrees in internal temperature. As it moves through space, it throws off flares of burning matter, developing voids in its surface called "sunspots." These disturbances affect our weather and long-distance radio reception, demonstrating the very large influence exerted over us by the sun, even though it seems to be so far away.

However, to return to the question of our sun's lifespan, it is estimated that our sun will burn for billions of years more. The estimate is based on the knowledge that it will take that long for the entire mass of hydrogen nuclei in the sun to form helium. After that, with its hydrogen depleted, the sun will be burned out, no longer able to produce energy. The sun's burning will supply our planet with energy for a very long time yet, for about another six billion years.

How Much of the Solar Energy Radiated from the Sun Reaches the Earth?

Only a very small fraction of the sun's total radiation is intercepted by earth.

The sun is a hundred billion billion megawatt generator. Of course, the earth catches only a fraction of the sun's radiant energy, and we can see from Figure 9.3 that this fraction is small. We can calculate just what this

---

*Of course, no sun goes on forever. But the times with which we are concerned in discussing stars are so long that it seems equivalent to "forever." Thus, the total life of our star, the sun, will be about 11 billion years, and its present age is about 5 billion years.

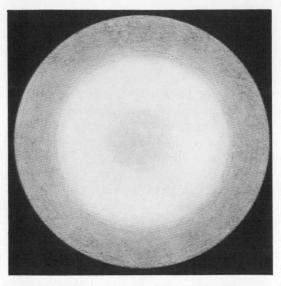

FIGURE 9. 2. The sun's disk. The thermonuclear reactions which produce the light and heat are occurring inside, where the temperature is many millions of degrees. At the surface, the sun is cooler, about 6000 °C.

fraction is using the fact that the sun radiates uniformly in all directions. A mathematical analysis shows that this fraction is about one-fifth of a millionth of a percent. This is a *very* small fraction. Would this part of the sun's energy be enough for our planet's needs, if we could collect, say, 1 % of it and convert it to useful energy media such as hydrogen and electricity? Before calculating our energy needs, let us see how much solar energy we can reasonably expect to be able to convert to some useful fuel (raw solar energy can only be used as a heat or light source).

SOLAR ENERGY REACHING THE EARTH

We saw above that the sun radiates at the rate of one hundred billion billion megawatts. The portion of this which the earth catches is one-fifth of a millionth of a percent, so the amount of megawatts which the whole earth receives from the sun is about two hundred billion megawatts.

This does not mean much until we know the amount of energy we need and compare it with how much we could get from the sun. Let us work out, first of all, the amount of energy received from the sun by the entire earth at the present time, per citizen, and then, after that, compare this with the energy available from the sun with the equivalent energy needs of the average person.

FIGURE 9.3. The sun radiates in all directions.
A very small swathe of this radiation is intercepted by the earth.

SOLAR ENERGY PER PERSON

The population of the earth was about four billion in the 1970s (and is
growing tragically rapidly, not as much due to the lack of the contraceptive
pill, as of the contraceptive will). The megawatts reaching the earth from
the sun are some two hundred billion. Thus, we see that we have about one
hundred megawatts, or one hundred thousand kilowatts, per person.

One-tenth of a million kilowatts per person is really a *very* large
amount of energy, 10,000 times more than the most energy-demanding so-

ciety needs per person. However, the comparison is not yet a fair one because we mentioned the *total* amount of solar energy reaching the earth, and we certainly would not be able to collect more than a small fraction of this in a useful form.

In any case, we can attempt a calculation of how much solar energy we can collect and convert to useable energy with optimism, since we do have this colossal amount to work with. We may still be comfortable even if we can collect only a tiny fraction of the total amount of radiation which strikes the earth.

How Much Energy Does the Average Person Consume?

The amount of energy which is needed in the world can be calculated as the sum of the energies needed by the various peoples, grouped geographically. Of course, different groups of people—the Asians, Europeans, Africans, North and South Americans—use fairly different amounts of energy per person because of the differing degrees of industrialization, and/or affluence, of the group. Industrialization is more prevalent in North America and Europe than in Asia and Africa. We have seen in Figure 2.1 that the average income per person in a country is proportional to the energy per person. Indeed, the amount of energy per person may be called the *comfort factor* in the area of the world concerned. It is as though energy circulating through a country's economy is the basis of its money, or that the money people have represents the energy of that community.

Therefore, to find out how much energy is needed for the world's population at present, we shall have to find the amounts of energy which the average person in each area is using. In Table 9.1, we list the kilowatts effectively required by the average person, according to data available on

TABLE 9.1. NUMBER OF KILOWATTS PER PERSON FOR VARIOUS GEORGRAPHICAL GROUPS

North America	8.3
Caribbean America	0.9
Other America	0.5
Europe	2.77
Western Asia	0.48
Far East	0.30
Oceania (including Australia and New Zealand)	3.31
Africa	0.27
Other countries	0.33
World average	1.46

the transportation systems, industry, and household energy consumption in the groups of countries concerned. If we total all the various uses in a country, each second in the day, 365 days per year, divided by the total number of citizens, we get the average amount of energy used per person in that country per unit time. The figure includes energy used not only by the person directly, but also the energy used by industry and government, etc., all of which contribute to the individual's standard of living. That is what the figures in Table 9.1 represent.

The world average is 1.46 kilowatts per person, or five billion kilowatts for the total population of the earth.

Trying to Allow for the Future: How Much Energy Will Be Needed by the Years 2000 and 2050?

Predictions of the future are uncertain, especially for futures of half a century away, in which new generations will be active. For short times, such as a decade or two, predictions can be made more accurately, based on assumptions of a continuation of the lines along which we have been going for some time. Regarding our energy situation, we are concerned with the line of the expansion of demand with time (Figure 4.4).

However, the reliability of this extrapolation will hopefully deteriorate rapidly, because the more industrialized countries have been expanding at very high rates which cannot be maintained if we are to survive our energy crisis. The unexpectedly high expansion rates have already brought on an earlier onset of energy exhaustion than had been predicted. Nevertheless, let us assume that we *can* reasonably estimate the amount of energy we shall need by the year 2000 by extrapolating from our present growth expansion curve (Figure 4.4).

As shown above, the total amount of energy we need now is about five billion kilowatts. Assuming that the growth of the world's energy needs will follow the growth pattern of the energy needs of affluent countries, in the year 2000 we will need about 2.5 times more, or about 15 million megawatts (15 billion kilowatts).

We could estimate how much energy we shall need in the year 2050 by extrapolating further along the line used above. But the answer we would get is much less reliable than that for the year 2000, for it is likely that the energy difficulties which we have been discussing in this book will give rise to a decreasing expansion rate (either voluntary or forced by circumstance). Towards the year 2000, a slowing down (i.e., a perhaps very unwelcome contraction) with resultant economic depression, unemployment, and a decline in living standards is more or less unavoidable (due to the slow development and acceptance of the necessity for building inexhaustible, clean energy sources).

We shall be content, then, to say that by the year 2000 the world will need about ten million megawatts and to see if we could obtain this from solar energy.

Solar Energy Is Dilute

Not more than about 0.02% of the sun's energy which strikes the earth will be convertible to useful energy.

The main difficulty with solar energy is that it is dilute. There is not much energy *per unit area* striking the earth. To get enough of the sun's radiation, a lot of territory must be covered with collectors. That is why the collection has to be in a far off region where the land is desert and costs very little.

In an oil-burning electricity plant, the amount of energy which can be produced is around tens of thousands of megawatts per square kilometer of land occupied by the plant. The amount of energy that can be collected from solar energy is, at best, one-tenth of a kilowatt per square kilometer. This diluteness is the most serious drawback of solar energy as an energy source.

The diluteness of solar energy per square meter of earth, even in sunny climates, is a problem.

How Much of the Earth's Surface Can Be Used for Solar Collectors?

We have shown above that the solar energy falling on the earth is about 200 billion megawatts. The portion of this which we can expect to collect in useful form depends on how much of the earth can be covered with solar collectors without upsetting and distrubing the environment.

Of course, the answer is going to be somewhat arbitrary—it depends on all manner of things. For example, what is the price of land in the country concerned? How much sunlight falls on the unit area of the land in question per year? One would not expect a good result from setting up large solar collecting areas in the Midlands of England, where most of the land has been covered with murky industrial undertakings for more than 100 years, and the sun appears only occasionally through the smog. On the other hand, in the southwestern U.S., central and western Australia, and in the great deserts of North Africa and Saudi Arabia, it would be possible to cover a large fraction of the land with solar collectors which would go essentially unnoticed if placed in remote areas.

A consequence of this situation is that most of the collected solar energy will have to be transported. Consider, for example, Great Britian. Such a densely populated, highly industrialized place, in a region of the Northern Hemisphere which is insolated at about one-sixth of the rate of solar energy arrival per unit area in, say, Australia, will *not* be the place where solar energy will be *collected*. The collection for European countries will have to be in the deserts of North Africa, for example. Then, the collected energy could be used to produce hydrogen, which can be sent to Europe in pipes.*

Let us take the bull by the horns and suggest that in the future, probably around the year 2020, we could cover as much as 1% of the earth's land area with solar collectors. In some sunny desert areas, it might have to be 50%, and in many areas, zero. The land area of the earth covers about one-fifth of the total surface area, so we are talking about covering 0.2% of the total area of the earth, including the seas, with solar collectors (see Figure 9.4).

What Will Be the Efficiency of Collection of Solar Energy?

Not only are we able to cover only some 0.2% of the earth's surface with solar collectors, but we also have to accept the limitation that we can only convert a fairly meager fraction of the light energy we receive from the sun to useful energy, per unit area.

How much will that fraction be? Here, again, a precise answer isn't known yet because it depends on details of the methods used. We shall be explaining these in Chapter 10. Let us take the fraction as 10%, a reasonable estimate of the efficiency for conversion of solar energy to electricity.

We have said that at most we will be able to cover a mere 0.2% of the earth's surface with solar collectors. This means that 0.2% of the solar energy striking the earth could be, in practice, collected and turned into useful energy. The total amount of energy reaching the earth from the sun is 200 billion megawatts, so this 0.02% corresponds to 40 million megawatts.

---

*This does, of course, unfortunately imply a political situation similar to that existing for oil today, in which not all countries can rely on their own land and resources to allow energy self-sufficiency.

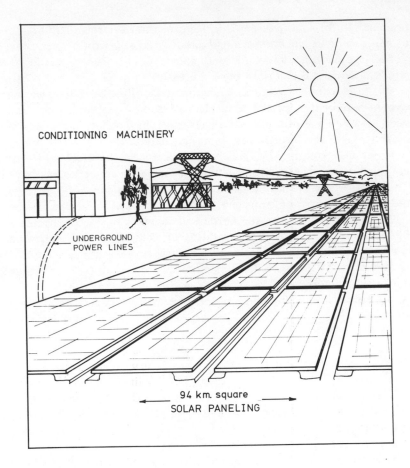

CONDITIONING MACHINERY

UNDERGROUND
POWER LINES

94 km. square
SOLAR PANELING

FIGURE 9.4. A solar farm.

WILL THE AMOUNT OF SOLAR ENERGY WHICH WE COULD COLLECT BE ENOUGH TO SUPPLY OUR TOTAL ENERGY NEEDS?

Let us do a calculation to see whether this is possible. By the year 2000, we said we are likely to have a total energy need of 10 million megawatts. The amount of solar energy which we can reasonably collect and turn into useful energy will be 40 million megawatts. Therefore, we conclude that we could obtain enough energy from the sun with acceptable collecting efficiency if 1% of the land area is covered by solar collectors, to support everybody on earth at a 2000 A.D. level of affluence. In fact, we should have quite a bit left over and be able to increase living standards in the underdeveloped countries to some extent.

Shall We Need More Energy than We Have Calculated Above in the Further Future?

The answer to this question is, probably, yes. It depends, however, upon one important aspect of the future, which we have avoided talking about so far, because the subject gets under peoples' skins and arouses hostile feelings. We are referring to the number of people on the earth—the earth's population. This is soaring upwards at an ever increasing rate. This is terrible news from the point of view of the supply of energy, as well as food. Unless we can reverse this trend, whatever we do to replace the fossil fuel supply with clean inexhaustible supplies from the sun will eventually be useless. Nothing can prevent our eventual collapse if people persist in producing so many new people. Our planet simply cannot sustain unlimited population, and famine and warfare are the most probable results. In all likelihood, a war will not simply reduce our numbers back to a supportable size, but may also damage survivors genetically and harm our environment to the point where it will not support life.

Thus, we do not know exactly how much energy we will need in the future because we do not know how large the world population will be and *therefore* how much total energy we will need. We do not know what living standards will be like in the future in the various continents. Were we to design a materialist's Utopia, we should have everybody living at a high living standard, with energy for heating and cooling our work and living areas, independent transportation for each individual, etc. This is a far-off goal, and we will not be able to achieve it for many decades, perhaps not even in our grandchildren's time. At the moment, therefore, let us concentrate on the more realizable goal of trying to attain a world supply of solar energy at the level needed for our world by the year 2000, and hope that we can slow down population growth before that.

To What Medium Shall We Convert Solar Energy for Use?

A discussion of solar energy is of little value without mentioning what we shall do with it and in what form solar energy will be used, when available. When it is first collected, solar energy will be either in the form of heat or electricity, and the form it will take depends on the method chosen for collection.

If we collect it onto mirrors which reflect their heat upon a central power tower (Figure 9.5), then the energy will first be in the form of heat.*

---

*This heat is used to vaporize water and the steam power that it makes will turn turbines, which turn the generators, which generate electrical energy.

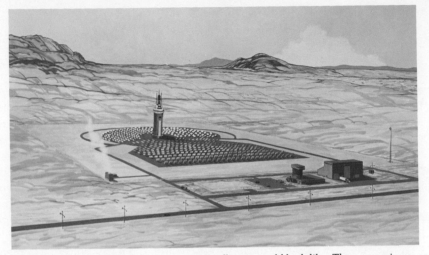

FIGURE 9.5. Drawing of what a "power tower" collector would look like. The many mirrors on the desert floor attract the sun and concentrate its light on a small area on the top of the tower. Here, steam is produced.

On the other hand, we might collect solar energy on photovoltaic cels, sketched out in Figure 10.4, which will convert sunshine directly to electricity. The manner of collection of sunshine will hence determine the energy medium, i.e., the fuel, to be used. Professor A. E. Hildebrandt is an authority on the power tower method (Figure 9.6).

If it is first produced as electricity from photovoltaic cells, then so long as the user areas are relatively near the production sites, we can send the electricity along wires to the areas where it will be used. It can be received in the towns from the electric grids, and supply our houses with lighting, heating, air conditioning, and power for appliances. Further, the energy from an electric grid could be used to charge batteries, run cars, and run factories. Along this line of development, we would have to rebuild a number of the machines which presently run on oil and natural gas.

Another possibility, which is discussed further in Chapter 11, is to use the source of electricity, or the electricity we get from solar heat, to electrolyze water to hydrogen and oxygen gases. The hydrogen gas would be piped to houses and factories, as is now done with natural gas to work internal and external combustion engines. With this latter scheme, we could run cars on hydrogen. The hydrogen would run into the car's cylinders, like the vapors of gasoline now do, be sparked off to explode in the cylinders, push pistons which work transmissions and turn wheels. See Figure 12.5.

FIGURE 9.6. Professor A. F. Hildebrandt, University of Houston, is a prominent figure in the development of the power tower method.

How Will We Get Solar Energy from the Places Where It Is Easily Available to Where It Is Needed?

If we look at a solar map of the world, we see that intense solar energy is found in regions of the world not far from the equator (i.e., within some 3000 kilometers, and better south than north of it). In this equatorial region are the *high solar energy areas* of the world, including places like North Africa, Saudi Arabia, parts of India, the Phillipines, northern and western Australia, and parts of South Africa and South America.

The most highly industrialized areas, however, are in different parts of the world. They are in the United States, especially on the East Coast and in the Midwest; in Europe, particularly in northern France, Germany, and England; in Russia, in the Moscow and Leningrad regions; and in two of the three Japanese Islands, Honshu and Kyushu.

These places are all large distances from the equator, generally some 4000–5000 kilometers away from it. Thus, if we are going to send solar energy from places where it is so abundant to places where it is needed, we will have to send it over distances of around 4500 kilometers (about 3000 miles).

Collecting vast amounts of energy 4000–5000 kilometers from where it is to be used seems like a fair-sized problem, and there are two ways of dealing with it:

1. Collecting solar energy nearer to the industrial-user area;
2. Utilizing an as yet undescribed method of transporting solar energy from the sunny areas to the areas where it is needed.

Another idea would be to gradually move the industry closer to the equator, but this would require extensive population migrations.

The first solution turns out not to be acceptable because the industrialized parts of the world are not as consistently sunny, due to weather and to smog from industry itself, and are not usually close to a cheap, sun-drenched desert area.

It would be better to contemplate the second possibility, and see if it would be economical to send solar energy over large distances. Then, we could use the sunny areas of the world, at the present mostly desert wastelands, to obtain the needed energy for the industrialized world.

It is not obvious how we could send solar energy 4500 kilometers. In fact, at first, the concept of "bottled" solar energy seems to be an unlikely proposition. We would first have to convert the raw solar (radiative) energy to some other *medium of energy*. One convenient medium is hydrogen. Solar energy would be converted into electricity, then the electricity used to electrolyze sea water, separating the water into oxygen and hydrogen gases. One could let the oxygen go back into the air or collect it for use in fuel cells, etc., and send the hydrogen through pipes* over long distances at high pressures, from the production sites (maybe on some of the Caribbean Islands) to places where the energy is needed, such as the Delaware River Valley industrial region of the U.S. There, the hydrogen could be used as we now use natural gas or gasoline.

It turns out, after cost calculations, that sending hydrogen in pipes over long distances is not an expensive process. The hydrogen fuel, arriving at its destination, would be more expensive than at the origin by about 0.2¢ per 1000 kilometers of passage for each kilowatt hour in fuel use.

How Much Time Will Be Needed to Make the Solar Collectors?

We have seen that sufficient energy for the world in the year 2000 could be given by the sun from the points of view of amounts of energy obtained, efficiency of conversion, and a reasonable land area covered with collectors.†

But what we can not be sure of is whether the governments of the world will act *in time* to have research, development, building, and commercialization completed within the 25 years or so before world oil and

---

*Over long sea distances, it may be more economical to convert the gaseous hydrogen to a liquid, and send this liquid fuel in tankers.
†Collectors on the sea may be part of our plans for the future.

natural gas supplies are exhausted (i.e., when shortages make it impossible for diminishing supplies to keep up with growing demand, and prices become prohibitively high).

Is it too late for us to hope for a solar alternative? Do we have enough time to put into motion the machinery necessary to cover such a vast area of earth with solar collectors? At present, it seems we cannot build solar collecting and distributing systems before the oil and natural gas run out, because building such worldwide systems of mirrors, wind collectors, seasolar plants (Chapter 10), etc. would require an emergency economy—special taxes, all kinds of government controls, and some sacrifice of personal freedoms in the countries which have based very successful systems, and hence their affluence, on freedom-loving enterprise (and very cheap energy from fossil fuels).

People do not welcome controls of the degree necessary to build the new energy systems in time. It would be like wartime, without a war, and people would not feel the same patriotic urge to agree to austerity. It is difficult to ask people to sacrifice parts of their own lives to achieve a dream for the future generations and avoid a disaster that seems remote in time ("After all, I may be run over by a truck tomorrow"). In addition, many trusted scientists and authorities do not foresee a true energy disaster, and these are the voices that are easy to believe. Governments of democratic states don't usually stress negative situations because there is always an opposition party ready to take over, offering positive pictures of the same situations if the government in power brings up unpleasant news, or cuts down the temporary pleasures (i.e., living standard) people have in a given election period. For this reason, people at large do not have too much opportunity of learning about the actual prospects in the energy situation. They do not have the information needed to decide whether they want to hasten toward building an energy system on earth, which obtains energy from inexhaustible, clean sources, or go on until the present exhaustible sources become so expensive that life in its present form is no longer possible. Then, the necessary research, development, and building would have become too expensive, and we should have to dismantle the civilization which we have built on the basis of finding the fossil fuels* (see Figure 9.7).

---

*A fanciful analogy can be drawn. It is perhaps as though finding the fossil fuels and using them was like finding the match to light the furnace, our civilization. This civilization has flickered along for a few decades, supplying about one-third of the world with the "good life." It gives us the larger brand needed to light the bigger fire of the fusion reactor, or the solar energy collecting system. If we do not light that fire before the present flare putters out, there will be no more technological civilization.

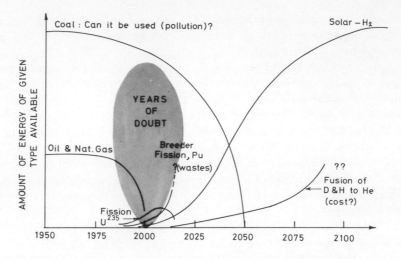

FIGURE 9.7. Oil and natural gas will exhaust around the end of this century. Can we build solar sources in time?

FIGURE 9.8. Typical barren Australian desert —useless except as a collecting area for the solar energy which made it so barren.

IF SOLAR ENERGY IS READILY AVAILABLE, WHY WASN'T ITS COLLECTION DEVELOPED MANY YEARS AGO?

The technology of solar energy was not developed earlier because it was thought atomic energy would replace energy from fossil fuels.

The answer is that there was no visible need to find out how to use it. There seemed to be so much oil. It was realized by very few that the expansion of the affluent and high-energy consuming world would use up oil and natural gas so fast that massive collection of solar energy would become important in a time (2–3 decades) of the same order as that needed for the development of a new technology. When anyone rocked the boat sufficiently to observe that one day the oil would run out, maybe even by the end of this century, there was always general laughter. "Cry wolf," people said. "We've heard about oil running out before, but they always find more."

The serious-minded stated confidently that atomic energy and breeder reactors would take over "when the oil does run out." People did not admit till the 1970s that atomic breeder reactor development (see Chapter 8) was immersed in problems whose solution might be too expensive for the society concerned, a society which would be getting poorer due to energy exhaustion.

A lesson to be learned from all this is "a good general plans his retreat," in other words, be ready with an alternative. A second lesson is that resources exhaust *suddenly* (Figure 4.3).

THE STRANGE SITUATION OF COUNTRIES WITH SOLAR ENERGY THAT DO NOT COLLECT THE SOLAR ENERGY

Countries where solar energy is plentiful but is not being developed and utilized include Morocco, Algeria, Libya, Egypt, and Saudi Arabia. An especially noteworthy example of a country which has barely stirred a finger to exploit its solar wealth is Australia, where about 50 times more money is spent on atomic energy developments than in developing the enormous solar energy that makes most of Australia an uninhabitable desert (Figures 9.8 and 3.8). Some countries in South American, e.g., Ecuador, Colombia, and Peru, concentrate their efforts on selling coffee and other natural products while ignoring the great potential wealth of energy pouring down upon them from the sun. Perhaps their citizens could be as wealthy as oil-rich Kuwait citizens if their governments would learn to collect, store, and export (in the form of hydrogen) their solar energy to the relatively sun-poor industrialized countries of the Northern Hemisphere.

Evidently, the recognition of solar power has not yet occurred in the minds of the legislators of the countries mentioned. Knowledge about the availability of solar energy needs to be spread. It is a matter of communication and realization. There was little knowledge to be discussed before 1973, and it is only in recent years that it has been realized that solar energy could be exported in the form of hydrogen, thus making countries which have plenty of solar energy wealthy.*

When the legislators realize this, it seems likely that they will invest the country's money into research and development aimed at building machinery to collect and sell the solar energy which their geographic position makes plentiful, and which the work of researchers could make practical.

---

*These statements apply to Australia more than any other country. Here is potentially the third largest solar collector in the world. The population has an educational standard equal to that of the U.S. or the U.K. The collection of solar energy could be done, and the Australian continent's tiny number of citizens (12 million) could be like the rich citizens of the oil-exporting states within a generation.

10

Converting Solar Energy to Useful Fuel

INTRODUCTION

The conversion of light radiation to heat energy is a matter of common ex-peience—things placed in the sun get warm. However, different materials absorb different amounts of heat. For example, something which is dark gets warmer than something which is light in color. The reason is simple. Black objects absorb light, so that no light is reflected from black surfaces to our eyes. The light, once absorbed, is converted to energy of motion, caus-ing the atoms and molecules in the object to move faster. This increased movement causes friction within the object, and is observed by us as a rise in the temperature of the object. In other words, the energy of the light is converted to heat energy, and the substance gets hotter (see Figure 10.1).

In contrast, a white object reflects light, which can be detected by our eyes. Since the light is reflected, it cannot be absorbed and converted to heat. Hence, white substances do not get hot in sunlight (this is the reason why people tend to wear white clothes in warm weather).

THE SOLAR SPECTRUM

White light, the light which comes from the sun, is actually composed of many different energies (i.e., colors that can be separated using a prism). Figure 10.2 is a graph of the intensities of the solar spectrum. At a certain energy, or wavelength, there is a maximum in intensity, in other words, most of the light has this particular energy. On either side of this maximum there is a sloping off of intensity. What we need, then, is a converter which collects light in the high-intensity range of the sun. Of course, the light

Plate Collector, or Panel, receiving radiant energy, converts it to heat energy, i.e. becomes hot.

Beam of Solar Radiation made up of Photons.

FIGURE 10.1. Radiant energy streams from the sun, and if a collector is put in its path it will become warm, especially if it is dark in color.

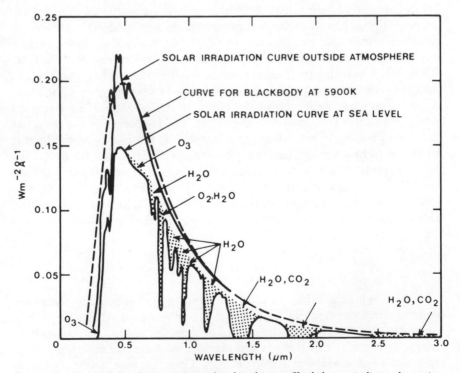

FIGURE 10.2. Spectral distribution curves related to the sun. Shaded areas indicate absorption, at sea level, due to the atmospheric constituents shown.

wavelengths which have a large amount of energy are more important for energy collection than that which has low energy. If our collector cannot be sensitive to all wavelengths, but only to those in a certain range, we should choose one which is more sensitive to light with an energy towards the left of Figure 10.2 (high energy).

Let us look now at the details of some of the methods by which we can convert solar energy to useable energy. We can get heat directly from sunlight by concentrating its energy on boilers containing water, which is then converted to steam. Or we can get electricity by letting solar energy fall upon a special crystal called a *photovoltaic cell*, as described below. Researchers in this area are shown in Figure 10.3.

THE PHOTOVOLTAIC METHOD OF CONVERTING THE SUN'S ENERGY TO USABLE ENERGY ON EARTH

When light arrives at a collector surface, a crucial question is whether or not it gets absorbed in the material being used as the absorber. Certain crystals can absorb light and produce a voltage (potential difference) and are thus called "photovoltaic."

A photovoltaic crystal (silicon is one type of such a crystal) will absorb and hence convert some energies of the light. Just which energies depends on a property of the crystal called the *energy gap*. If the energy of the light is greater than this energy gap of the photovoltaic crystal, then the light will be absorbed, and the energy of the light converted to electrical energy. For example, the light can be used to lift electrons in the crystal, energetically speaking, from a low energy to a higher energy, within the solid framework of the crystal.

Suppose then that we consider light which comes from the sun, and which is greater than the energy gap in silicon crystal. This light will be absorbed in the crystal. Our crystal will contain two sections of silicon crystal which have different properties, and a junction where the two meet. Inside the crystal, the light's energy is transferred to electrons on one side of the junction only, elevating the energy level of these electrons. The energized electrons dash around inside the crystal section, and some of them arrive at the junction between the two types of silicon. The electrons can not cross the junction, and a buildup of electrons at one side of the junction results.

The junction has on the other side a countercharge, or net positive charge. A *photovoltaic couple* (Figure 10.4) has been formed. We could also say that a solar cell, or a solar battery, has been formed, because the two sides of the silicon in the photovoltaic couple are acting in effect like the two terminals in a battery and can produce a current. This can be measured by

FIGURE 10.3. The discoveries for the conversion of solar energy to electricity are more than 1% efficient. This photograph shows Pearson, Chapman, and Fuller of the Bell Telephone Laboratories.

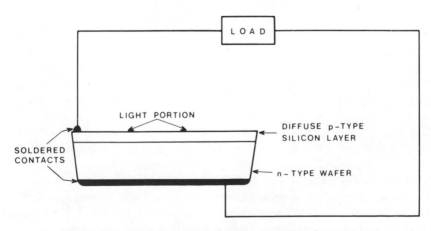

FIGURE 10.4. Mechanism of the function of a photovoltaic generator.

taking two wires and attaching one to each of the two parts of the photovoltaic crystal (Figure 10.4).

Each of the silicon photovoltaic crystals produces about 0.8 volts. The voltages needed for practical applications are much larger, at least 100 volts or more. But we can create these larger voltages by stringing several cells together, with the positive terminal of one cell joined to the negative terminal of the next, and so on. For example, to get 100 volts, it would be necessary to have 100/0.8 = 125 cells joined together, positive to negative, so that the voltages add together.

Then, to get a current of any desired amount, one must connect the couples arranged to give the appropriate voltage. To do this, one joins negative pole to negative pole of groups of cells. One would make up a series of photovoltaic couples connected $+ - + - + -$, etc., so that the right voltage was attained, and then take this multicouple and connect it to another multicouple by joining negative-negative. The coupling of the opposite poles increases voltage, and the similar pole coupling, current.

How much current can be obtained from such coupling? This will depend upon the *area* of the couple. Let us say it is about one square meter, because the greater the area, the more light will arrive and be absorbed. Let us take the best condition, with the sun directly overhead and no clouds. In this case, we will get something in the region of 100 amps of current from a square meter of couple.

Is the Photovoltaic Method of Collecting Solar Energy Too Expensive?

For a long time, from about 1954 until about 1974, scientists thought that the photovoltaic method would be far too expensive for practical use. The reason was that the photovoltaic crystals of silicon are difficult to manufacture, involving much labor and machinery, and therefore very expensive. Consequently, the electricity made by the crystals from the sun would also be too expensive for practical use by the public.

One problem is that the silicon used for photovoltaics has to be pure, or else the higher energy electrons in it are "destroyed," i.e., lose their energy and become deactivated when they strike the impurities in crystals. Another requirement is that the silicon exist only as single crystals. The usual form of crystalline material is polycrystal, which consists of many billions of crystals held together in one big structure.* The boundaries between each of the minicrystals, or *grains*, are called grain boundaries, and when the electrons moving about in a polycrystal strike the boundaries they lose energy and deactivate. Such electrons do not make it to the junction be-

---

*The parts of these crystals are extremely small, about a thousandth of a millimeter across.

tween the two silicon sections of Figure 10.4, and will therefore not contribute to the voltage of the cell.

To stop this energy-losing deactivation, we must make single crystals (crystals having no minicrystals, and therefore no deactivating boundaries), and this means making large sheets of silicon which are single crystals of a size, say, one meter square and one millimeter thick. Making single crystals is difficult and expensive, and until 1972 it seemed unlikely that they would ever be made at a price which would allow generation of solar electricity from silicon photovoltaic cells at an acceptable cost. Then, a method was found in 1972 which enables these pure, single silicon crystals to be made

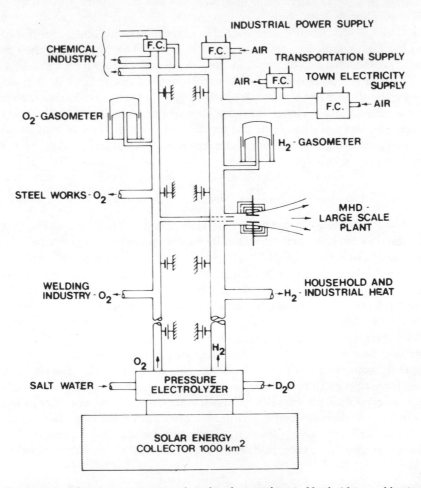

FIGURE 10.5. Schematic presentation of a solar plant working in North Africa and having an area of about 1000 km². Hydrogen would be transmitted at 50 atm.

FIGURE 10.6. A strip of the photovoltaic material, cadmium sulfide. (Courtesy of Westinghouse Corporation.)

much more cheaply, as long as large batches are involved. This may provide us with electricity at a price in the region of 5¢ per kilowatt hour from this method. The goal is to get 1¢ per kilowatt hour for large-scale industrial electricity, while it is still produced from oil and coal. Figure 10.5 is a schematic of a solar plant.

Thin Film Photovoltaics

Another approach to the conversion of solar energy to electricity is to use a "thin film photovoltaic." It requires a thickness of about one-thousandth of a centimeter of silicon to absorb the light. There are other types of photovoltaics, the most well known of which is cadmium sulfide,* for which the thickness needed to absorb the light is only a hundred-thousandth of a centimeter. Light is absorbed within this distance (i.e., causes electrons to be active in the photovoltaic), and the distance is small so the electrons have only a short path to go to reach the collector. Impurities can be present to a greater extent in cadmium sulfide than in silicon because the cadmium sulfide is so thin that there is much less time for the electrons to meet an impurity as they pass through the crystal. This means that such crystals need less purification and are therefore cheaper to make.

In 1979, the estimated cost of a cadmium sulfide photovoltaic cell with all the equipment to go along with it was about $500 per peak kW (estimated by Brody and Shirland at Westinghouse), providing electricity at about 2¢ per kilowatt hour. A strip of cadmium sulfide is shown in Figure 10.6.

*It is, in fact, a dual couple of cadmium sulfide and copper sulfide.

Photovoltaic Collectors in Orbit

It has been suggested that it would be advantageous to place the collectors for photovoltaic conversion in orbit around the earth, because there would then be no interruption of the collection of radiation in desert areas by clouds and at night. Figure 10.7 shows a photograph of a model of this idea. The collected light is converted to dc electricity, and then this is power-conditioned, i.e., converted to ac at a frequency in the megacycle range. In this condition, the power could be *beamed* back as microwaves to collecting stations on earth, and these would be far smaller in area than the very large areas of desert which we should need for the collection of solar energy on earth.

Moreover, there would be no need for transport of the energy over long distances, from desert areas to the energy-using regions, because the sun would always shine on the orbiting collectors, and they could beam back energy anywhere on earth (See Figure 10.8).

This may all sound too good to be true, and so it is, at present. The platforms, even if there were many hundreds of them, would each have to be very large (football fields in size) and, hence, very heavy. There would never be any question of boosting a whole platform into orbit—it would

FIGURE 10.7. A photograph of a model of an orbiting solar collector. The two panels contain the photovoltaic material, and the power conditioning and beam transmission are in the middle.

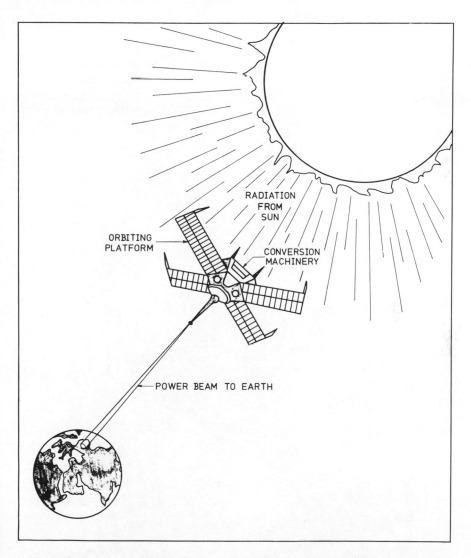

FIGURE 10.8. Solar collection from an orbiting platform. Energy is collected on photovoltaic cells. The electrical energy is converted on a board to electricity and beamed to Earth, where it is re-collected and stored as hydrogen to be distributed to the cities.

have to be built in space, using a space shuttle. The difficulty is cost. At present, it costs about $10,000 per kilogram to put something into orbit. If and when the cost is reduced to several hundred dollars per kilogram, then orbiting solar collectors will be built.

THE MIRROR CONCENTRATOR METHOD

This method is shown in Figure 10.9. It uses an array of mirrors which are connected to motors and timers to move and switch them on and off in

FIGURE 10.9. A square-mile mirror array with boiler atop a 1500-ft tower.

small time intervals. The timers which drive the motors are coupled to information on the apparent position of the sun. Each mirror has its position adjusted every few minutes so that all the mirrors accurately reflect the sunlight onto a tower which is placed several hundred meters above the several thousand mirrors. Each mirror must move independently, because each mirror has a different position on the ground relative to the tower.

The reflectors have an additional task. They reflect the sun, of course, but they also focus it on the top of the central tower. Now, whatever is at the top of the tower is going to get hot indeed, because it is receiving, by the arrangement which is shown in Figure 10.9, the whole of the energy falling on the area of the mirrors.

Suppose that we have a square kilometer (one million square meters) of area. If the sun were directly overhead, there would be a maximum of one million kilowatts shining down on this square kilometer. So, if there were no energy loss, we would be putting a million kilowatts (1,000 megawatts) onto the power tower.

Of course, losses will occur, say 50%, and there will be a further loss in going from the heat power arriving at the top of the tower to electric power, say about 66%. Thus, the net electrical production from the tower could be about one-sixth (160 megawatts) of the solar energy striking the mirrors.

This method has one fairly large disadvantage compared with the photovoltaic array method. It depends upon the sun actually *shining*, i.e., the solar disk being visible. The lenslike parts of the mirrors have to have something to focus on. That is not true for the photovoltaic method or for the ocean thermal gradient method (see the next section). The first of these can use both clouded sunlight and heat, the second heat only. Neither depends on the sun actually shining.

Ocean Thermal Energy Collectors (OTEC)

A different idea for obtaining a useful form of solar energy is the *ocean thermal gradient method* (Figure 10.10). In the hotter regions of the world, there are areas in the oceans where the surface temperature of the water is 25 °C or more. In the depths of such oceans, however, at a kilometer or more beneath the surface, the waters are cooler, around 5 °C.

Let us consider what would happen if we place an apparatus with a boiler on top and a condenser underneath the boiler (Figure 10.11) in the ocean, so that the boiler is in the hot surface waters and the condenser underneath it. A tube extending for 1000 meters into the cold depths is used to pump water from the depths to the condenser. If we place a liquid which boils at less than 25 °C in a boiler on the top, hot layer of the tropical ocean,

Figure 10.10. A model of an ocean thermal energy collector. (Courtesy of ERDA.)

it will boil and become a vapor. If we then pass the vapors into a condenser suitably cooled by water from the depths, the vapor will be cooled down towards 5 °C, and if its boiling point is above this, it will condense to a liquid.

We need to find a liquid, therefore, which boils at less than 25 °C but higher than 5 °C. Then, this liquid will boil when it is in a boiler on the surface of the tropical sea. When we circulate the vapors, after they have worked a heat engine, to the cold condenser, we shall get the vapor back as a liquid again. One such liquid is ammonia, which will boil in the required temperature range at high enough pressures, and is easy to obtain.

The sequence of events is shown in Figure 10.11. Ammonia is evaporated

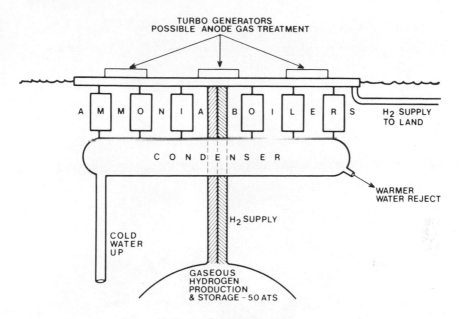

FIGURE 10.11. A solar-sea power plant.

and is jetted onto the blades of a turbine. The turbine blades turn and drive an electric generator. The ammonia vapor is passed to the condenser in Figure 10.11, which returns it to the liquid state. This liquid can be put back into the boiler, heated up again by the water on the surface, evaporated again, pushing the turbine blades around again, and so on. The electricity can be used to produce hydrogen by electrolysis.

This, then, is a method of making useful fuel (hydrogen), taking the energy from the hot surface water of the oceans. This energy, which the water obtained from the sun, is heat energy, so that it, too, is solar-produced energy. In fact, in this case, our oceans are serving the role of huge, natural solar collectors.

To What Extent Have the Methods Described in this Chapter Actually Been Built and Used?

Very unfortunately, of the three methods we have described, only the first one (the photovoltaic) has actually been put into practice (Figure 10.12), being used in space vehicles. The other two methods are in the pilot

FIGURE 10.12. An actual photograph of a space-orbiting laboratory supplied with about 25 kW of electric power from the silicon photovoltaic panel seen from above and at the side of the vessel. (Courtesy of NASA.)

plant stage,* and the first multimegawatt production will occur in the early 1980s.

It is indeed sad and wasteful that we have built so few pilot plants of the mirror concentrator method and the ocean thermal gradient method as yet. Wasteful, because the energy from the sun already exists. We do not need to worry about starting energy-producing reactions or containing them, as in nuclear plants on earth. All we need to develop is the machinery for *harnessing* the energy. We can build solar energy collectors ourselves, and can use natural solar energy collectors like tropical oceans. The problems are not fundamental scientific questions, but only difficulties in engin-

---

*In France, a large mirror concentrator has been made and operates successfully. But it is a vertical one and used to concentrate sunlight on very small areas, the size of a coffee cup, to make high temperatures, e.g., 3000 °C. Barstow, California, will have a 10-megawatt solar thermal mirror concentrating plant in 1982.

eering and construction. There is no concern with unestimable danger factors such as possible health hazards or radiation accidents. Until a satisfactory, permanent method for energy production, such as atomic fusion, has been developed fully and proven safe, we must look toward the safest, easiest renewable energy sources available. Our greatest hope is the sun.

11

Household Energy
from the Sun

The part played by energy in the household (heating of water, home air-conditioning, cooking, lighting, and refrigeration) comprises about 25% of the energy load in industrialized countries like the U.S. It would therefore be very worthwhile to get this energy directly from the sun by collecting it on rooftops, at least in the sunnier parts of the world. A pioneer in this field is seen in Figure 11.1.

There is already in the 1970s a small-scale version of this going on with respect to heating of household water, particularly in Israel and some of the former French colonies in Africa, where high solar irradiation makes the situation very favorable. The heating of water for use in the kitchen and bathroom takes up about one-quarter of the household energy, which is itself about one-quarter of the whole energy load; so 6–7% of the total national energy demand is for domestic hot water. Were we able to shift that burden to rooftop solar water heaters, it would be a good beginning for the general use of solar energy.

In this chapter we are going to describe how solar energy could be developed to give us the *whole* of the household energy load, including lighting and cooling. The simplest part of our energy technology, and one in which solar energy could replace oil and coal even now, is hot water in the kitchen and bathroom. Such solar apparatus has reached the commercialization stage, i.e., can be bought by the public. The rest of the necessary equipment is in the pilot plant stage of research and development, except for the production of household electricity, which is between the research stage and the pilot plant development stage.

We shall take a look at each of these possibilities in the following sections.

141

FIGURE 11.1. Maria Telkes, a pioneer in solar energy research.

FIGURE 11.2. The Australian Government Research organization has had for many years an activity to develop solar panels for water heating.

The Production of Hot Water

A good solar heater needs a substance which absorbs the maximum amount of light impinging on it. Therefore, a black absorber should be used since black absorbs all the light reaching it, reflecting none. Black panels placed on a rooftop absorb the available sunlight. The energy from that sunlight is absorbed by molecules of the black panel substance. This absorbed energy causes increased motion of the molecules and manifests itself as heat, i.e., the temperature of the panel increases.

Consequently, to set up a solar hot water heater, a house rooftop contains panels of black material, which absorb heat in the presence of sunlight. Water from the main water supply (or a supply collected from rain water) flows through pipes touching the black panels and absorbs the heat which has come from solar irradiation (see Figure 11.2).

As heated water has a lower density than cold water, the water from the hot panels rises up to a storage tank kept on an elevated part of the roof. It is recirculated over the collecting panel of black material on the roof. Hot water is thus available during sunlight hours and for several hours afterwards. The mechanisms for these events are shown in Figure 11.3.

A rooftop water heater will supply a *fraction* of the hot water needs for the household, depending upon that part of the world in which one is situated, and the local amount of sunshine. In a country like Australia, even in the coldest part of the country (South Australia), the fraction of household water heated by the sun is 70–80% of total need.

Space Heating of Houses

By space heating, one means heating the air inside the house, at present accomplished by fossil fuel methods—oil or gas coming from pipes or by oil heating air and then fan-forcing the heated air through ducts into each room, or by the inefficient method of electric resistance heaters. The inefficiency of the latter is seen as follows. The fuel from which our heat comes at present is oil, coal, or natural gas. If we burn oil, say, in a basement furnace, and duct the heat to rooms, we are getting a high proportion, perhaps some 80%, of the heat into the household air. If, however, we go to electric heating, the oil is burned at the electricity generating station, creating electricity at an efficiency of about 40% whilst 60% is thrown away (escapes to the surrounding air through the chimney). Then, we send 40% converted as electricity to homes and *it* is used in resistance heating. The advantage, of course, is completely clean heat. If the basic fuel is cheap enough, the inefficiency of electrical resistance heating may be compensated by its conven-

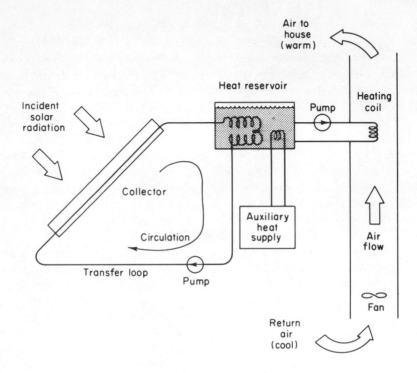

FIGURE 11.3. Schematic diagram of a simple home solar heating system.

ience. However, we have already pointed out the expense and eventual exhaustion of the basic fossil fuels used. We *cannot* afford to waste them.

Solar energy can provide space heating, and the principle of how this
works is not too different from that of hot water heating. Air circulates underneath the black solar collection panels. The air absorbs the heat from the
panels exposed to sunlight, and is then circulated around the house by being
pumped through ducts (see Figure 11.4).

We also need a storage device for the heat energy (the house must still
be heated at night), and this device may cause the main expense of space
heating. There are several approaches to storage of solar-generated heat,
and we are going to give more information on them in Chapter 16. For one
house, we could have a hole dug in the garden, lower into it a 10,000 gallon
tank, fill it with water, and run the roof-heated air into the water, heating
it. In this way, the heated water in the tank heats the cold air circulated
through it at night from the house. As the cold air is heated by the warm
water, it would be pumped into a duct system to rooms in the house.

Alternatively, we could use the warmer water of the "eutectic latent heat storers," which will be described in Chapter 16. The apparatus containing this would be kept in a basement storage facility of fairly large size. Heat from air circulating under the hot panels on the roof is pumped through a mixture of molten salts in the basement container, imparting heat to it, and subsequently melting the salts which compose it. Later, when no more warm air is available from the roof (i.e., in the evening), warm air for the house would be obtained by passing *cold* air from the house through the hot, molten salt. This would heat the circulating air, which would then be fan-forced around the house.

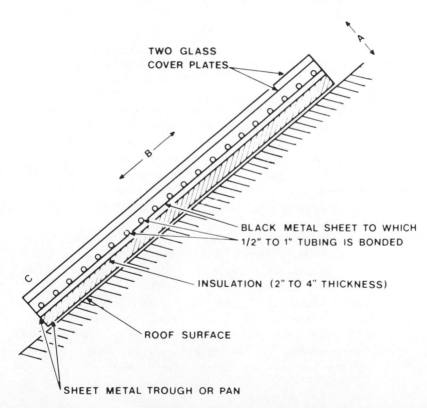

FIGURE 11.4. Diagramatic sketch of one possible system (elevation section) for a solar collector for residential heating and cooling. *Notes:* The ends of the tubes are manifolded together; there are one to three glass covers depending on conditions. *Dimensions:* Thickness (a direction), 3–6 in.; length (B direction), 4–20 ft.; width (C direction), 10–50 ft.; the slope is dependent on location and on winter–summer load comparison.

SPACE COOLING

The cooling of the air in hot climates contributes much to the living standard. When people are uncomfortably hot they do not work well, and have less ability to perform energy-demanding tasks.

Several methods of using solar energy to provide space cooling are possible. One of them is to simply use electricity which can be made from solar energy, by means described in Chapter 10, to operate a refrigeration system placed in the roof of a house, cooling air and fan-forcing it through ducts around the house. However, the direct production of electricity from solar energy is still in the research and pilot-plant stage, while other methods of getting household solar energy are in the development and even the commercialization stages.

Another idea for getting space cooling is possible:

Space cooling could be carried out with solar energy in such a way that the stronger the sun shines, the cooler the house becomes.

The principle here is that evaporating liquids cool the vessel in which they are stored. Why is this? It is because evaporation is a process which consumes energy, the energy of heat. When a substance evaporates to form a vapor, it is the heat of the vessel which causes vaporization of the liquid. The transfer of this heat energy results in the cooling of the vessel donating the heat, which we call air-conditioning or refrigeration. Therefore, if we blow warm air over this container, that air will be cooled, giving some of its heat to the cool container.

In Figure 11.5 there is a vessel on the left side, which contains a solution of the substance ammonium chloride, dissolved in liquid ammonia. When a salt is dissolved in a liquid (ammonium chloride in ammonia in this case), the pressure exerted by the pure liquid is lessened.

Now, imagine what would happen if we arranged a period of time (in the range of minutes) in which the left-hand vessel, containing the ammonium chloride in liquid ammonia, were irradiated with a beam of solar energy, and thus made hot. The heating will cause ammonia to evaporate from it, and this gaseous ammonia will pass over into the right-hand vessel. The right-hand vessel has not been irradiated, and is therefore colder than the left-hand vessel. As the pressure from the hot left side forces the ammonia vapor into the cold right-hand vessel, the vapor condenses back to a liquid.

Now, we arrange a second cycle, lasting a few minutes. The solar energy which was shining on the left-hand vessel is "turned off" (a timer mechanism closes a shutter), so that the left-hand vessel no longer gets heated.

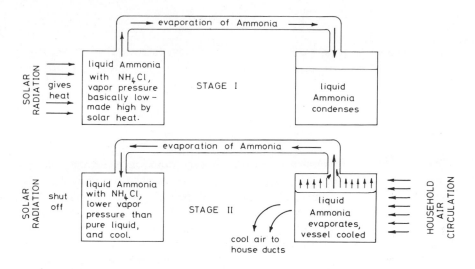

FIGURE 11.5. Schematic for solar cooling. Solar radiation is periodically applied to the container of liquid ammonia containing ammonium chloride in solution.

Ammonia is no longer evaporated toward the right-hand vessel, which by now contains liquid ammonia under the pressure built up when the lefthand vessel was hot.

Thus, the second cycle proceeds as follows: the heat source (the sun) is no longer applied to the left-hand vessel; the liquid ammonia in the right-hand vessel is no longer under pressure from the evaporating ammonia in the formerly heated left-hand vessel; as the liquid is no longer pressed down by the pressure of incoming ammonia, it *evaporates,* and therefore *cools the vessel* which contains it.

So, the right-hand vessel is cool, and all we have to do now is pass warm air from the house over the vessel, or over a heat-transfer device of large area kept cool by the right-hand vessel. The household air, therefore, loses heat, and can be pumped in ducts back and forth between the house and the source of the cold. The cycle of letting the sun heat the left-hand vessel, and letting ammonia evaporate from the right-hand vessel, can be continued when the sun has set, or has been obscured by clouds.

This, then, is the principle of how we could have a house *cooled* by solar heat. In fact, the hotter the sun, the greater the degree of cooling. The energy provided by the sun can indeed accomplish many tasks, the variety of which is demonstrated by the number of solar alternatives, not only for heating, but for cooling as well.

Household Electricity

In the last chapter, we talked about methods of converting solar energy to heat and electricity, describing three methods: the photovoltaic production of electricity directly from the absorbtion of sunlight, which produces electricity in solar cells; the direct concentration of light from mirrors focused on a tower to produce steam which drives turbines; and the collecting of energy on a large scale from sea-solar gradients.

Insofar as houses are concerned, either of the first three methods could be used. There are many experimental solar houses in existence (Figures 11.6 and 11.7) with photovoltaic panels installed on the roof which convert sunlight directly to electricity to supply the house's electricity for lighting and household appliances, e.g., television, radio, or kitchen appliances.

Another method, the "power tower" method, might be used. Here, we would have four or five houses arranged so that optical lenses on their roofs all focus the sun's rays onto a tower. The tower would have a water tank which uses the concentrated heat to produce steam. This steam would then work a turbine and electric generators, producing local electric power for groups of houses. In the future, we may build houses in a circle, around a

FIGURE 11.6. Professor K. Boer outside the house "Solar I," built at the University of Delaware. Some of the panels on the roof contain cadmium sulfide panels which convert incoming radiation to electricity.

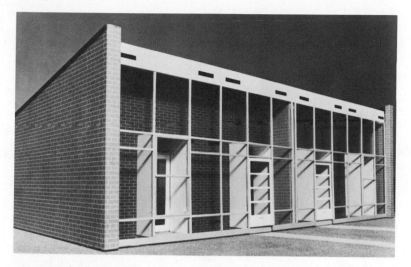

FIGURE 11.7. A solar house constructed by the University of York in the U.K. Note the large roof sloping to face the sun.

power tower, and have lenses and collectors on rooftops of the houses, so that there is an increased amount of energy for the power tower (absorbing panels, boilers, turbines, generators) to produce substantial amounts of electricity.

This type of solar energy provision for houses—conversion to electricity—is less advanced in engineering development than the other schemes we have discussed. It seems at present that it can be done, using photovoltaic couples (as discussed in Chapter 10), although the price of such electricity may not be at a competitive level until there is mass production of the photovoltaic devices on a large scale.

Electricity for houses could be supplied by photovoltaic couples.

WHAT DO WE DO WHEN THE SUN GOES DOWN?

Even in the deserts of Saudi Arabia and Australia there are dark, cloudy days, and the world experiences an average of 12 hours of darkness per day. Providing energy storage so that energy can be available during the night and cloudy weather is an important part of solar heating.

There are several storage devices we can use. Large bodies of water are good heat storers and transfer agents. We can also use high-temperature liquids—sometimes called molten salts—and these absorb and store heat when they melt and release it again when they freeze. If the rooftop panels

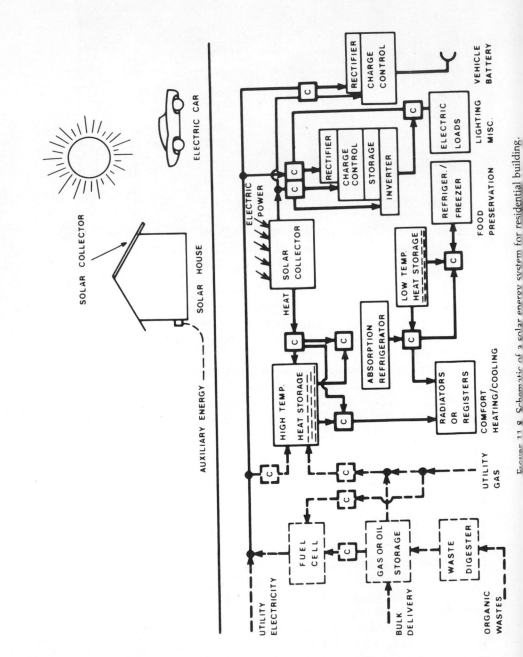

FIGURE 11.8. Schematic of a solar energy system for residential building.

are photovoltaic and produce electricity, we can use storage batteries, charge them during daylight hours, and use their electric power when needed during the night. Finally, if we could use solar electricity to electrolyze water, which would yield hydrogen, we could store the hydrogen in pressure tanks beneath the earth, or make the hydrogen into a more easily stored substance which would release hydrogen when needed (for example, a hydride such as iron-titanium hydride could be used). This would lead the way toward *solar-hydrogen* transportation and export for areas of the world where there is sunlight in abundance.

WILL SOLAR ENERGY FOR HOUSEHOLDS BE COMMERCIALLY AVAILABLE BEFORE OIL RUNS OUT?

We have said, in the introduction to this chapter, that rooftop solar water heaters are being sold now in the warmer parts of the world, and are used to provide hot-water systems in houses. However, installation still has to be organized individually by each customer—no service system exists—and the customer has to find his own maintenance person and advertisements and listings are scarce. Energy companies are hardly willing to help the householder gain his independence from the oil, gas, and growing nuclear plants from which the companies make their profits. Figure 11.8 shows a schematic of a solar energy system for a residence.

Solar-energized housing could be a new mass-produced industry, like automobile manufacture, by the year 2000.

The development of the rest of the technology of providing houses with solar energy (apart from water heating) is approaching commercialization. People are forming companies and getting into the technological and commercial stage of development (Chapter 5).

How long will it be before we can *buy* a house run entirely on solar energy? This depends upon how much of the people's money (taxes) is allocated for research in solar devices and, later, for their development. If we made an all-out effort and spent large amounts of money on developing this source—perhaps \$25 per person in the whole country per year—then the necessary research and development might be done in about 10 years. After that, allow about 10 more years for commercialization, mass production of plants, and marketing.

How will citizens react to the rising price of fuel oil? Revolution? Constructively? Passively? It depends upon the degree to which the news gets to the citizens that alternative heating and energy possibilities to the expensive fossil fuels exist, waiting to be developed and advertised for general use.

12

Transport and Industry Run on Electricity and Hydrogen

INTRODUCTION

We have observed in Chapter 11 that it would save about 25% of our energy load if all the homes in an affluent, industrial country were run on solar energy, instead of fossil fuels. Solar-based hot water alone would save 6% of the energy load, in parts of the country where it could be used. However, the major part of our energy needs are outside the home. In fact, over half of our energy goes into the activities of transportation and industry.

Private transportation, the ideal that every family, and every person, have an individual vehicle for her use alone, is an important component of what is called "freedom" in democratic and consumer-oriented societies. These are the societies which have developed according to the principle of allowing people as much freedom as they can purchase. A car, and all the individual traveling and commuting it makes possible, is available at a price. Only recently has the realization begun to surface that we may not be able to fuel these cars *at any price.*

This faith in never-ending fossil fuel supplies is evidenced by our increasingly car-dependent life styles. Many newer urban and suburban communities are constructed so that a car is essential for bringing children to schools, getting to work, shopping, and recreation and social life. Many neighborhoods do not even have sidewalks, making walking dangerous and unpleasant, especially for children. Public transportation systems are successful only in certain large cities in which rush-hour traffic and parking trouble have simply made car travel impossible.

The gasoline shortages of the past few years have opened the eyes of the public to its dependence on gasoline, an idea which is stressed by oil companies, but the exhaustion (in the lifetimes of many of us) of *all the* fossil fuels on earth is deemphasized. Short-term shortages, always temporary, raise prices and profits for oil companies. But a true realization of the limited supply, and the fact that we will not always find more oil, might lead to a search for alternatives (e.g., an electric car) right now. This would not be welcomed by the oil industry nor by the automotive industry because of the implications in respect to re-tooling costs.

Nevertheless, we are going to run out of oil, natural gas, and thus gasoline, in some few years. It is essential that we see how other sources of energy, apart from the nuclear ones (see Chapter 13), could be used to run transportation and industry.

Running Cars in the Post-Fossil-Fuel World

There are two main ideas of how cars could be run in the post-fossil-fuel society. The most well-known alternative (Figure 12.1) is the electric car. Many people assume this means that the car is run on batteries. How-

Figure 12.1. Multipurpose vehicle designed by Transportation Systems Laboratory, Anderson Power Products, Inc.

FIGURE 12.2. The Kordesch Car, an Austin-America, driven directly off of lead–acid batteries with hydrogen cylinders driving hydrogen–oxygen fuel cells, which in turn drive the batteries. Top speed, 80 km.p.h.; range; 300 km; peak power provided by batteries, 20 kW; average power provided by fuel cells, 6 kW. (Courtesy of Union Carbide.)

ever, there is another possibility for electric cars, using an electrochemical device called a fuel cell, as seen in Figure 12.2, described later in this chapter.

There is another way in which we could run our cars—a solar-hydrogen way. This would use hydrogen in internal combustion engines (see Table 12.1). At present, we explode gasoline with oxygen in internal combustion engines, but few engines design changes would be needed if we were to use hydrogen instead of gasoline. It would be much cheaper and more

TABLE 12.1. COMPARATIVE FUEL PROPERTIES

Property	Hydrogen	Other fuel
Heating value, BTU/lb	53,000	20,000 (gasoline)
Minimum ignition temperature, °F	1,065	1,000 (butane)
Theoretical flame temperature in air, °F	3,887	3,615 (butane)
Flammability limits, % by volume in air	4.0–74.2	1.9–8.6 (butane)
Maximum flame velocity, ft/sec	9.3	1.03 (butane)
Specific volume liquid, liters/kg	14.3	1.33 (gasoline)
Energy density: BTU/ft³	2.50×10^5	9.38×10^5 (gasoline)

convenient for the car manufacturers to adjust to a hydrogen fuel than to convert to electric battery cars.

Could We Run Cars on Batteries, Charged by Electricity?

The answer to this is yes, but there are many other questions which arise as to the practicality of this idea. For example, what sort of performance would we get from an electric car run on batteries? Is 60 kilometers per hour (37.3 miles per hour) and 60 kilometers between recharges satisfactory for a car predominantly used in cities (about three-quarters of all driving is "in-town" driving)? First, let us examine two types of batteries which could be used in electric cars.

Cars Run on Batteries Which Work with Lead Electrodes

Figure 12.3 shows a lead–acid battery, which uses lead electrodes. This is the battery most frequently used at present, and indeed it is used for starting cars which run on gasoline. When electricity is put into this battery it

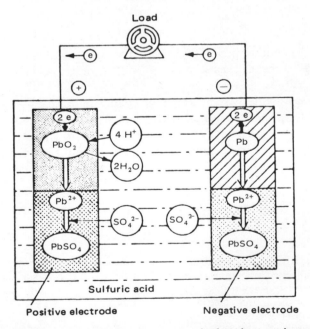

FIGURE 12.3. The discharge processes in a lead–acid storage battery.

"charges up," i.e., it stores electrical energy by forming lead on one elec-trode, and lead oxide on the other. If there is no charging current, we can connect the terminals of the battery across the terminals of an electric motor. During this "discharge," electric current will flow through the wires in the motor and do work, i.e., energy flows. So, the electricity which is put into the battery on charging is stored in the form of lead and lead oxide, and can be made to flow into some device needing electrical energy.

One drawback of the lead–acid battery is that it is made of lead, so it is *heavy* per unit of electrical energy which it can store. Due to weight, we cannot load a car with many lead–acid batteries, which limits the distance the car can go before recharging is necessary. If we put enough batteries on board to make it go about 400 kilometers, the car would be so heavy that it would have difficulty in pulling itself. A car can only go a short distance on one charge of a lead–acid battery, around 60 kilometers. However, this is the sort of distance which most people travel on a daily basis, for commut-ing to work, etc., and so a lead–acid battery would be acceptable in these cases.

Another important question is whether we have enough lead to supply all the cars in the United States with lead-acid batteries. Probably not. If we had a big increase in demand for lead, especially if it were known to the seller that it was in high demand, he would raise the price, making lead less accessible and the cars too expensive to run. Alternatively, if the demand for lead were satisfied, a shortage would develop, since the world's supply of lead is relatively small. For this reason, as well as the limitations on range, a lead–acid battery would not be best for the first commercial, mass-produced electric car.

The Sodium–Sulfur Battery

A battery which runs on sodium and sulfur can also be used for run-ning cars. The battery is shown in Figure 12.4. It still needs more research and development to make it stable and long-lasting, but a car could go five times as far on sodium–sulfur batteries as on an equal weight of lead–acid batteries. It would travel 250 kilometers or so without having to have the batteries changed for fresh ones.* This would require a few changes in con-sumer habits.

The maximum speed of a car driven by a sodium–sulfur cell is not es-

---

*In about one minute at the recharge station. It is a myth to think that battery-powered cars would have to wait around for hours having their batteries recharged.

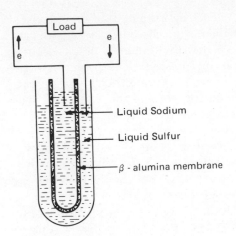

— Liquid Sodium

— Liquid Sulfur

— β - alumina membrane

FIGURE 12.4. Schematic diagram of a sodium-sulfur cell.

tablished yet. Of course, it could be any speed a fossil-fueled car could do, but the faster it goes, the more energy used per kilometer, and the smaller the number of kilometers traveled per recharge. Figure 12.5 shows two researchers in this area.

A sodium–sulfur battery, with a few more years of development and commercialization, could give performance equal to that of cars driven by the polluting, exhaustible fossil fuels.

An unfortunate feature of the sodium–sulfur cell is that it has to work at a high temperature, about 360 °C, and the cell would need plenty of insulation. The sulfur would solidify if left for more than a week without use. To start up again, an auxiliary battery could be used to melt a pathway through the sulfur, and the heat which would then be generated from the battery itself would do the rest.

A Source of Energy to Charge Batteries for Electric Cars

Any source of energy would do. Solar-generated energy is not yet developed enough to assume the burden of transportation. Until hydrogen fuel from solar converters is conveniently accessible, one possible source for charging electric car batteries is coal. Coal resources could last into the next century, longer than oil and natural gas, and there is no doubt we could use coal for making electricty to run cars. This would be a temporary measure, only for the short period between when we run out of oil and natural gas and when we have built enough solar converters to produce sufficient hydrogen to run cars on this completely clean fuel.

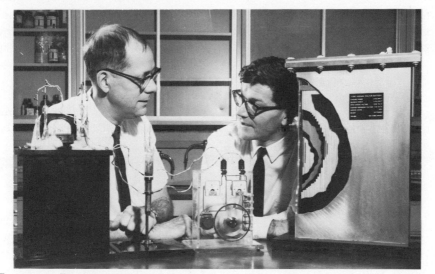

FIGURE 12.5. Drs. Kummer (left) and Weber (right), the Ford Motor Company's electrochemists, involved in the development of the new sodium–sulfur battery, which can be used for running cars and trains. (Courtesy of Ford Motor Company.)

HYDROGEN-DRIVEN CARS

Hydrogen is a fuel which we could have in abundance from a solar-hydrogen economy. It could be used to power cars and other vehicles. Hydrogen could be introduced into the vehicle (Figure 12.6) as a gas, stored in high-pressure cylinders. Alternatively, it could be liquefied, but liquid hydrogen must be kept at a low temperature ($-250\,°C$), and this requires a specially insulated tank, which would be an added expenditure in the cost of

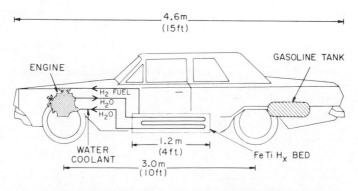

FIGURE 12.6. Multifuel vehicle (hydrogen–gasoline).

the car. Another approach to storage would be to use an iron–titanium metal alloy and charge it by letting it absorb hydrogen gas. Discharge of the hydrogen gas is brought about by changing the conditions so that the gas is desorbed from the iron–titanium.

Figure 12.6 illustrates the principles of a hydrogen-fueled car. The principle is essentially the same as for a gasoline car. Hydrogen is a gas at normal temperatures. It would be mixed with air in cylinders and exploded in the same way as gasoline, the force of the explosion causing the piston to be forced out the cylinder, etc.

HYDROGEN-DRIVEN PLANES

There is a chance for changing from gasoline to liquid hydrogen for running the next generation of airplanes. You can see in Table 12.2 that the weight requirement of liquid hydrogen is suitable for planes. In fact, it would need less weight of fuel per unit of distance covered than gasoline, and the plane could have a larger freight and passenger load. Each flight could therefore make more profit, and air transport should become more economical. Running planes may be one of the first applications of a technology in which we use hydrogen as a general fuel. An advantage of using hydrogen would be that aircraft would not pollute the atmosphere, as planes driven on hydrocarbon-type fuels do today.

There is a problem in utilizing hydrogen in aircraft. We could not take the present aircraft, fit them with low-temperature fuel tanks to hold liquid hydrogen instead of gasoline, and take off. Far from it. Although liquid hydrogen is much lighter than gasoline (per unit energy provided), it requires more space for the same amount of energy than gasoline. It takes 3.8 times as much volume for hydrogen as aviation fuel to go the same distance, and we would therefore need more room in the aircraft to hold the hydrogen. For airplanes, weight, not volume, is the critical factor; so our larger-volume hydrogen planes are still carrying less weight of fuel and can carry

TABLE 12.2. COMPARATIVE JET-PROPULSION AND LIQUID-HYDROGEN AIRCRAFT CHARACTERISTICS[a]

Fuel	Gross weight, lbs	Empty weight, lbs	Fuel Volume, Ft^3	Length, Ft
Jet propulsion	750,000	327,000	7,130	306
Liquid hydrogen	510,000	316,000	28,050	343

[a]Mach # 3; 3,500-nm range; 300 passengers.

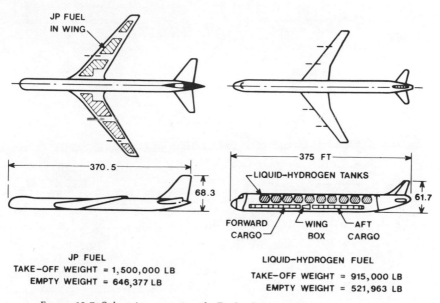

FIGURE 12.7. Subsonic cargo aircraft. Payload, 265,000 lbs; range, 5070 nm.

more cargo. A simulation is shown in Figures 12.7 and 12.8. So hydrogen is for the next generation of planes, the ones being designed now. (See also Figure 12.8.)

FUEL CELLS: HOW TO GET BACK ELECTRICITY FROM HYDROGEN DERIVED FROM THE ENERGY OF SOLAR RADIATION

The first practical fuel cell was made by Tom Bacon (seen in Figure 12.9) and Rex Watson at Cambridge in 1955. Currently, these fuel cells are being used in space vehicles. They supply the energy in these vehicles which allows astronauts to keep warm, have electric light, read in bed, and keep in touch with mission control by radio and television.

A fuel cell is an electrolyzer working in reverse (Figure 12.10). An electrolyzer uses electricity to produce hydrogen and oxygen from water. In a fuel cell, conversely, *electricity is created* when hydrogen and oxygen are combined. When a hydrogen molecule, composed of two hydrogen atoms bonded together, enters and contacts the electrode on the left side of the cell, it dissociates into the two atoms of hydrogen. These atoms dissolve in the solution, producing electrons (among other products). The electrons migrate around the electrical circuit, as shown in Figure 12.10, and then go

Figure 12.8. Alternative design of liquid-hydrogen subsonic passenger aircraft.

Figure 12.9. Tom Bacon of Cambridge University, first to build a practical fuel cell.

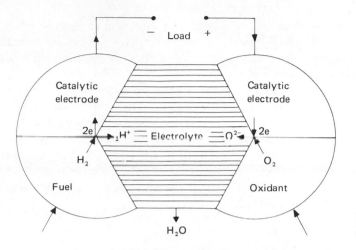

FIGURE 12.10. Schematic of a fuel cell: hydrogen reacting with oxygen.

through a resistance and do electrical work. They cause a potential difference to exist between the two electrodes, which can be used to provide useful work. For example, to run an electric motor in a car, or to provide the lighting in a house. During this passage, the electrons start out with much energy but end up with little. They give it up to overcoming an electrical resistance. If that resistance is the circuit of an electric motor, they will drive the motor.

Fuel cell cars are now in the pilot-plant stage. If used in conjunction with batteries (which supply a larger pulse of current for starting) they could probably be commercialized within a few years. But there has to be a *willingness* of the great international groups which make cars before such a commercialization could occur. Willingness with companies means the prospect of profit. When the companies can make a better profit making electric rather than other kinds of cars, they will make them. That is what will control the availability of clean electric cars (Figure 12.1). One way to encourage the necessary condition would be to make the tax on gasoline driven cars high, or that on electric cars low.

The Poor Efficiency of Ordinary Engines

When chemical fuels like oil or gasoline explode inside cylinders of internal combustion engines, only about one-quarter of the heart energy which is produced by the reaction actually gets converted to mechanical energy. If we are using the engine to make electricity, the next step will be the conversion of the mechanical energy to electrical energy, which is almost 100% efficient. Thus, overall, making electricity will be less than 25% efficient using internal combustion engines. Very large plants may get up to 39% efficiency, which is still low.

The basic reason why there is such poor conversion efficiency rests upon an idea called the Carnot efficiency. The name refers to a French engineer, Carnot, who, in the 18th century, deduced theoretically that the efficiency of the conversion of heat energy to mechanical energy by an engine depends on the temperature change in the engine during the conversion (i.e., the difference between the initial and final temperatures). The larger the temperature change, the higher the efficiency. For an average engine, the efficiency is around 30%.

The Better Efficiency of Electrochemical Engines

We have described how a fuel cell works. An electrical potential arises across the cell terminals by formation of water from hydrogen and oxygen electrochemically, rather than by combustion in an engine. Thus, electricity is produced *directly* in a fuel cell, and no further energy conversion steps are necessary.

What is the efficiency of the conversion of the chemical energy to electrical energy? The maximum, ideal efficiency turns out to be about 94%, which is three times greater than the efficiency of ordinary combustion engines. Unfortunately, a realistic estimate of the efficiency of a fuel cell is much lower than 94%. This is because some additional energy is needed to get the cell running initially, and this energy does not therefore contribute work toward running the load. It is spent in getting the electrons to enter the electrodes. This extra energy need reduces the fuel cell efficiency to a range of 50–70%. This is still about twice as efficient as the internal combustion engines. A schematic of the electron sink and source is shown in Figure 12.11. Thus, a fuel-cell-powered car which uses solar-generated hydrogen fuel is a very promising alternative to our fossil fuel vehicles.

Running Industry on Hydrogen

The story gets very complicated when it comes to running industry on hydrogen. It would be next to impossible to describe all the possibilities. At present, industry is fueled partly on natural gas, partly on gasoline and crude oil, and partly on electricity made from coal or oil. Sometime during the next few decades, only electricity from coal, or methane produced from

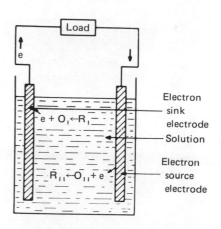

FIGURE 12.11. Schematic diagram of an electro-chemical energy producer.

coal, will still be available, *if* there is a sufficient building of coal conversion plants to make methane in time.

Hydrogen could replace natural gas and oil for heating and for driving internal combustion engines, and atmospheric pollution would be diminished. The hydrogen would be produced by the electrolysis of water using atomic or solar-produced electricity. Figure 12.12 shows a space vehicle fueled by liquid hydrogen. If solar energy were used to electrolyze water and so produce hydrogen, the latter would be piped or shipped from highly sunlit, equatorial areas of the world to industrialized areas, and there would be used to replace natural gas.

Would hydrogen be used in liquid form or gaseous form? Probably in the gaseous form, because it is cheaper to store than the liquid. In some instances, however, it might be worthwhile spending some extra energy to liquefy the hydrogen and keep it in tanks. Such low-temperature tanks are called cryogenic storage tanks, and are very carefully insulated. Also, we have mentioned storage in hydrides. Recently, a lot of progress has been made by the use of iron–titanium alloys, which are cheap and convenient.

FOODS FROM HYDROGEN

There are many chemical processes which use hydrogen. For example, hydrogen can react with nitrogen to produce ammonia. It could be used in production of organic-type compounds, and even in the production of food. This may seem far-fetched, but the trend is clear that food will be increasingly synthetic in years to come.

For food production, we need other substances in addition to hydro-

FIGURE 12.12. At the Kennedy Space Center in Florida, a space vehicle is shown in the pre-launch position. It is fueled by liquid hydrogen, and the tank containing this can be seen in the background. (Photo courtesy of NASA.)

gen. One useful substance which we have in abundance is carbon dioxide. We can get it by burning carbonate rocks, or by extracting it from our atmosphere (of which it comprises about four-hundredths of a percent, which is actually a very large amount).

With the carbon dioxide from the atmosphere and hydrogen from solar energy, there is a possibility of massive use of an important organic chemical reaction which produces a chemical, formaldehyde. Formaldehyde is an organic compound which could be the basis of synthetic foods. Such production processes are likely to be enzymatic, i.e., those which take place due to the presence of an enzyme.* Specifically, some enzymes cause the combination of nitrogen from the air with formaldehyde to form proteins.

Because our bodies live to a large extent on proteins (our bodies *are* about 14% protein), a protein which could be made from the carbon dioxide in the air, hydrogen formed by solar-powered electrolysis of water, and nitrogen from the air, could be a basic raw material for the future. One would not then continue to breed animals for the purpose of killing and eating them, a process which has a primitive aura, as well as being too energy inefficient for a densely populated planet. Already, a protein food is being made synthetically for cattle by the British Petroleum Company, starting from oil and using nitrogen from the air and an enzyme. Thus, foods have already been manufactured from hydrocarbons and air-derived nitrogen, and so could probably also be produced by the carbon dioxide–hydrogen route outlined above.

Fats and starches have not yet been produced from such simple compounds, but it is easy to foresee the synthesis of such foods as long as we have basic materials available, such as carbon dioxide, water, and nitrogen. These ideas depend, of course, on the presence of abundant water and solar energy, to make hydrogen via electrolysis of water.

METALLURGY

Many processes taking place in the metallurgy industry today using fossil fuels, e.g., burning coke and emitting polluting carbon monoxide as well as carbon dioxide, could also be done with hydrogen. One reason why the use of fossil fuels should be replaced by hydrogen (in addition to their hastening exhaustion) is that the waste products would be *clean: No dirty substances would be pushed into the atmosphere.* For example, we could form iron, the basic metal in steel, by reacting iron oxide with hydrogen. Iron made in this way, with solar-derived hydrogen, would be very pure, and the process is much cleaner than the present blast-furnace method used today.

---

*An enzyme is a naturally occurring substance which we can extract from bacteria and other natural substances, and which serves as a catalyst for reactions in solution.

13

Tides, Geothermal Heat, and the Big Winds

Wind generators, or "aerogenerators," are often called by their older name: windmills. These were first seen in the twelfth century, and have been used mainly to drive machinery for farming industry, i.e., grinding grains. Modern wind generators can be seen on farms today. The rotors have generators linked to them, often behind the rotor. Such a generator is small and produces 1 or 2 kW, even in fairly strong winds of 20 km/hr. The modern wind generator on a farm is used to charge batteries which are kept near its base, and these are used to supply the electrical needs inside the farm house at all times, independent of the momentary wind strength.

COULD WIND BE A RELIABLE SOURCE OF ENERGY ON A LARGE SCALE?

To the untrained eye, it would appear that winds are random and therefore unpredictable. However, recordings of wind over many years, taken at the same places, show that there is a predictable yearly average wind for a given locale, technically called the "mean annual wind." This is the wind velocity measured at short intervals (every hour or so) and averaged for the whole year. It was found that this mean annual wind is more or less constant, year after year, for a given place (see Figure 13.1).

So, as long as you choose a specific place, and no major changes occur in the environment (for example, no skyscrapers are built nearby to slow the wind, as they were between about 1875 and 1910 in the data of Figure 13.1) you can expect to have the same average wind velocity from year to

169

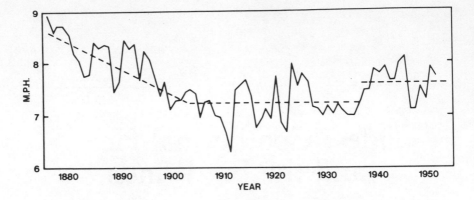

FIGURE 13.1. Wind annual mean velocities, as derived from Adelaide Weather Observatory data. The change in 1938 was caused by a change in position of the anemometer.

year. This means that we could consider using wind as a *reliable* source for massive energy collection. What mean annual velocity is required to make wind a useful energy source?

It is helpful to know the magnitudes of typical "mean annual winds." The mean annual wind velocity in a windy city might be 12 km/hr, but the mean annual wind on a nearby stretch of water 20 km out to sea may be 24 km/hr.

The wind velocity over the water is usually greater than the wind over nearby land. Wind over water does not encounter obstacles, e.g., hills or cities, as it does over land, so that it can gather greater momentum. Also, winds tend to blow faster at greater altitudes. At jet plane heights (10,000 m and more) the wind may be more than 200 km/hr.

Could Wind Generators Produce Household Electricity?

We can give some kind of answer to this if we know household energy demands. The daily electricity requirement of a household of four people in the United States is about 35 kW hrs.

Let us suppose that we had a 2-kW wind generator, i.e., if the wind is blowing right at the maximum speed which the generator can stand without operating a cut-out mechanism which prevents its going too fast, 2 kW will be produced. We can do a simple calculation assuming that the 2-kW generator will produce 48 kW of electricity per day. For the estimated 35 kW hr needed, the generator must run for about (35/48) 24 or 17½ hours to sufficiently charge the batteries.

The maximum wind which the generator can handle will not always blow. Let us assume that the mean annual wind is half the maximum wind velocity for which the wind generator is rated. One's first inclination is to suggest that to get the same energy one would need a doubling of the wind generator capacity. Unfortunately, it turns out that doubling is not enough.*

These are very rough estimates, but they give one the idea that a small generating plant could supply the electrical needs of the house *if* the mean annual winds are strong enough in the locale. Of course, as mentioned earlier, a storage battery system is needed along with the wind generators; so that electricity is available when needed, and not only when the wind blows.

More Wind Energy Estimates

It is interesting to see how much energy is supplied with wind speeds of 30 km/hr, supposing the engineers had built gigantic wind generators with a radius of 100 m, as shown in Figure 13.2. For example (Table 13.1), if the mean annual wind is 30 km/hr and the radius of the generator 100m, an average energy of 5 MW throughout the year can be produced. This means that production of energy at the rate of 5,000,000 W (or 5000 kW), is possible. This is a respectable amount of energy indeed, and sufficient to support about 500 people in an affluent, industrial country with all the associated transportation, industrial, defense, and household energy needs (and up to 5000 people in a less industrialized society).†

It is clear that the big winds would not be a negligible energy source. *And they are clean.*

The Big Winds

To understand the generation of the earth's winds, recall the principle that hot air rises and cold air falls (because cold air is denser than hot air).

---

*The reason is that wind energy is proportional to the velocity cubed:

$$\frac{\text{Wind energy for wind velocity } v_1}{\text{Wind energy for wind velocity } v_2} = \left(\frac{v_1^3}{v_2}\right)$$

Hence, if the maximum wind velocity at which the generator is allowed to function is 60 kph, i.e., halves, we shall need $(60/30)^3 = 8$ times the wind generating capacity which we used earlier to produce in 17½ hours the needs of the family house. At first sight, we seem to need a 16-kW generator, but in fact, the 35 kWhr is needed over 24 hours, so that around 12 kW of windmill power is needed.

†This single giant windmill could produce the equivalent of about 1 million gallons of gasoline per year.

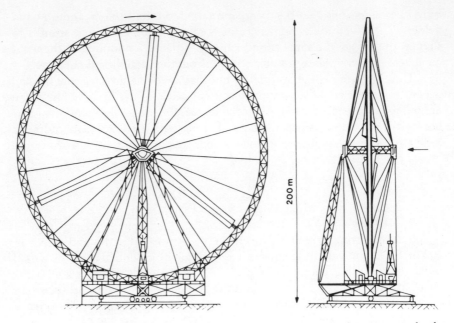

FIGURE 13.2. A design for a wind generator which does not bear weight on a central axle (Mullett, 1956).

TABLE 13.1. EXAMPLES OF THE WIND ENERGY EQUATION
FOR A ROTOR OF RADIUS 100 METERS[a]

Power of rotor (MW)	Mean velocity of wind during year (km/hr)
24.6	50
12.6	40
5.4	30
1.6	20
0.2	10

[a]Calculations have been made basically with the wind power equation $(16/27)\, \rho v^3$ per unit area. However, some efficiency and averaging factors have been allowed for.

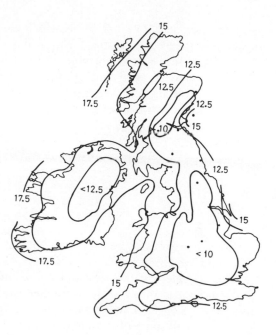

FIGURE 13.3. Wind map diagram. Large lines show the mean annual wind in the U.K., in miles per hour. Off the west coast, wind generators could be very productive, helping the energy supply.

Also, it is warmer at sea level than at higher altitudes, e.g., on top of a mountain (see Figure 13.3). This is because the earth stores the sun's heat better than the air of the atmosphere does.

> *In certain regions of the world there are wind belts in which the wind is stronger and more constant than in other places. Obviously, wind farms should be built in the wind belt regions of the world. Hydrogen would be the medium of transport of energy elsewhere.*

The equator is consistently warmer than other parts of the world, and the warmer air there rises particularly strongly. As it rises, it cools off, getting further away from the earth's surface. When it gets cold enough at greater heights, it begins to fall towards earth once more. However, the world is rotating, so as the air mass falls closer to the earth's surface, the air over the equator appears to have a west-to-east relative velocity (in the Northern Hemisphere), i.e., the falling air becomes a wind. The winds which sweep along the southern* oceans are a natural consequence of the presence of the air on the earth, the superior heating effect of the sun at the

---

*They are less observable near the ground in the Northern Hemisphere because they are broken up and reduced by the land masses.

equator, the gravitational force which pulls the heavier cold air down again, and the earth's rotation. They will go on in a 24-hr cycle for as long as the earth goes around the sun, and would only be altered if there were significant changes in the energy of the sun reaching the earth, whereupon life would fade anyway. We can rely, therefore, upon the world's major big winds, and know their velocity at sea at a given place, so long as we are talking of yearly averages.

Big winds could be one of our large energy sources, but they are at present used only in a minor way as farm generators.

Winds at Sea Are Stronger

One of earth's biggest winds is that which roars across the South Pacific, and which was called in the days of sail "the roaring forties." The number 40 refers to the most northerly latitude where the wind is found, and "roaring" to the noise the wind made among a sailing ship's rigging. Before the days of steam, when sailors set out from England to go to Australia, they went well south of the Cape of Good Hope at the bottom of Africa to get into this wind, whereupon their windjammers would zoom across the Indian Ocean at 14–16 knots (about 18 mph). In the early days of steam, the windjammers in this part of the ocean would be caught by the roaring forties wind (the wind itself averaging about 21 knots—25 mph) and often whip the early steam boats doing only 11–14 knots.

Winds near the surface of the sea are faster and stronger than those near the surface of land because they do not have obstructions on the sea to break them up and slow them down. The wind can gather momentum, unbroken by hills, scrub, or buildings, and the net velocity is about 1.5 times higher over sea than over land (see Figure 13.4).

Because wind force increases greatly with its velocity, we can get much more energy from sea winds than from the winds on land. For this reason, it would be well worth building wind farms (collections of floating platforms on which aerogenerators are mounted) on the sea.

Another place for collecting the big southern wind might be on mountain slopes (e.g., in Chile) facing the sea, and looking westward, because as the height increases, the mean annual wind velocity tends to increase, too. One such mountain slope wind collector is shown in Figure 13.5.

How Would Wind Energy Be Stored on a Massive Scale?

Just as solar energy varies every day, day-to-day wind velocities are irregular, too, and therefore also necessitate the storing of energy gathered from it. There are several ways to do this. As stated, the energy can be stored in batteries which charge when the wind is blowing and can be discharged into the household electrical system at any time. This is the method used on most farms.

If wind energy is to assume part of the burden now carried by fossil fuels, 10 or more wind generators might be built around an electrochemical hydrogen producer, placed on platforms floating on the sea. The electrolysis tower would produce hydrogen from sea water and hydrogen fuel would be stored in tanks beneath the sea at pressures of some 100 atmospheres (i.e., 100 times the pressure exerted by our atmosphere on earth), to be sent through pipes to land. One would have to have a big storage capacity since during summer months the winds average below demand and in the winter months above demand.

There is engineering and development research to be done here. We do

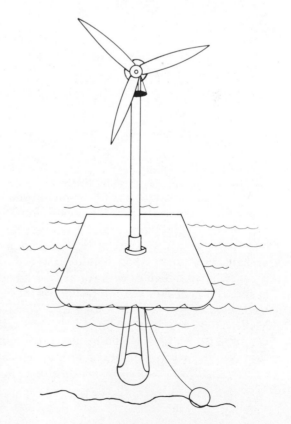

FIGURE 13.4. A sea-born wind generator. Winds on the seas are greater than those on land—note the counterweight under the generator. This would counteract the turning moment which the wind would exert on the generator, which could capsize it.

FIGURE 13.5. Mountain slope wind generator, "Grandpa's Knob."

not have experience in building large platforms upon which the rotors would sit, and we have already made the point that no rotors of the size we need have yet been made. The wind would exert a tremendous turning movement on the platforms and tend to capsize them. Platforms would have to be built with counterweights deep underneath them, like the keel underneath a yacht, which would prevent the platform from turning over.

The main difficulty with rotors which have a size as big as 100 meters in radius is that they are too heavy for their axles and we would have to have another way of supporting them. Perhaps we could support them (as Mullett suggested) by resting the weight on a dozen smaller gear wheels (see Figure 13.2) so that it would be distributed on twelve axles instead of one. However, even if we cannot build rotors, we can still break the rotor system up into small wind wheels which we already know how to build. Together, these would cover only slightly more area than the big rotors for the same amount of energy, as explained in Figure 13.6.

WOULD MASSIVE WIND POWER BE A PRACTICAL PROPOSITION?

The only sure way to answer this, of course, is through more research, including construction of an experimental giant rotor. Massive wind power has one great advantage: it does not require any new basic research and new discoveries (as *is* needed in atomic energy research, particularly fusion). Large-scale engineering should be able to provide the technology needed to build big wind power rotors. The associated large-scale electrolysis development has to be tackled, and hydrogen collection and storage technology built, together with means for transport of the fuel to land. Figure 13.7 shows Les Mullett, a pioneer in rotor design.

One troublesome aspect is that when we electrolyze the sea—mainly a solution of sodium chloride (common table salt) and water—we get not only the desired hydrogen at one electrode, but also oxygen mixed with chlorine gas at the other electrode. Chlorine is a poisonous, green, malodorous gas which would act as a pollutant in the surrounding atmosphere, worse than any of the nonradioactive pollutants from fossil fuels systems. However, methods can be envisaged whereby we can take this chlorine and convert it to a soluble species which could be sent back into the sea.* (See Figure 13.8.)

Could big wind generators be a source of large-scale energy for the future? Yes, they are a promising possibility for the near future, i.e., before the end of the oil supply approaches in the twenty-first century.

The difficulty of each method to provide energy is relative. Big wind generators are difficult to build compared with smaller ones. But compared with building environmentally safe atomic breeder reactors, or fusion machines, the task appears easy. Its attainment could certainly occur before the end of the fossil fuels, whereas the attainment of the controlled fusion reaction in a practical and economic form seems to be too far distant; so that we cannot rely on it as a replacement for oil.

The development work needed to make the generation of electricity and hydrogen from wind belts practical is far less than that necessary to make (for example) atomic fusion practical. Wind energy on a large scale could be realized before the exhaustion of oil and natural gas.

---

*Chlorine gas would be converted to hydrochloric acid via reaction with water, releasing oxygen into the atmosphere. The quantities of hydrochloric acid would not be drastic enough to alter the chemistry of the sea significantly.

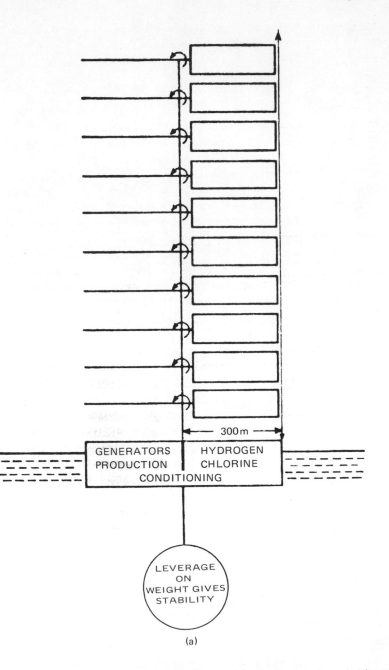

(a)

FIGURE 13.6. Possible designs for large sea-born generators. (a) Individual wind panels (aluminum frames and sail cloth) rotate central drive shaft, and then rotate themselves to a minimum profile position for rotation direction against wind.

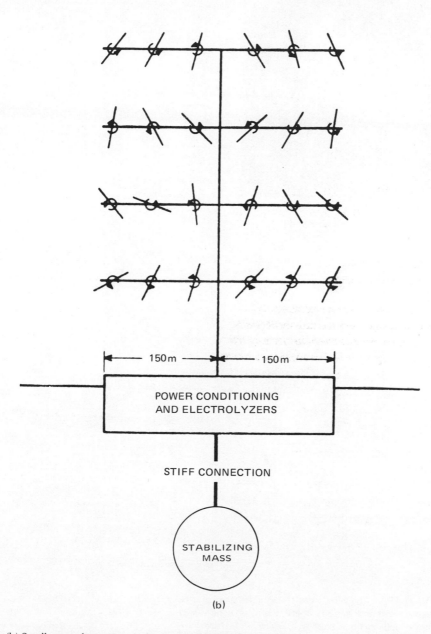

(b) Small rotor alternative. (A large rotor has a 25-m radius.) A large number of smaller rotors is used, with a motor generator attached to each.

FIGURE 13.7. Les Mullett, a pioneer of scale wind energy development.

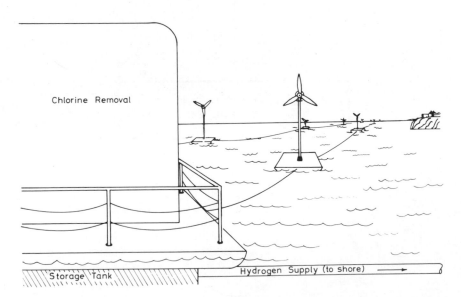

FIGURE 13.8. Wind generators turn, and the generator behind each produces electricity which is sent to a central station, where hydrogen is produced by electrolysis. Chlorine is also produced, and can be converted to oxygen and hydrochloric acid. The latter is returned to the sea and dispersed.

It Is Always the Cost that Counts

Scientists think of new ideas, engineers try to make them work, but finally they are only commercially acceptable, becoming attractive for people to buy, and therefore desirable for a company to produce and thus get a profit, if the cost is less than that of the competing possibilities. For example, we could have had solar energy using photovoltaic cells 20 or so years ago, but at that earlier time the cost of these cells was too high, and therefore the electricity produced by them too expensive.

It's a matter of cost with respect to wind power also. Not only must the actual electricity production far out at sea be considered. Transportation costs to get the energy to land and into the cities, are also significant.

Briefly, the expenses we must consider for wind energy are the following:

a. wind generators,
b. electrolysis plants to get the hydrogen from the sea,
c. a chlorine removal plant,
d. tanks to store the hydrogen,
e. pipes to take the hydrogen to cities,
f. fuel cells to convert it back to electricity again, when needed.

One estimate of all this is given in Table 13.2.

Energy from the Tides?

The coastal waters in most parts of the world rise and fall a few feet each day. Figure 13.9 shows the moon interacting gravitationally with the earth, causing these tides. If we can manage to trap the *risen* water behind a dam

TABLE 13.2. Cost Estimates for a Wind-based Hydrogen Energy System

Source of Expense	Dollars per kilowatt
Wind generators	200–600
Electricity generator	100–300
Hydrogen production	200
Undersea storage facility	100
Fuel cell	300–700
Total	900–1600

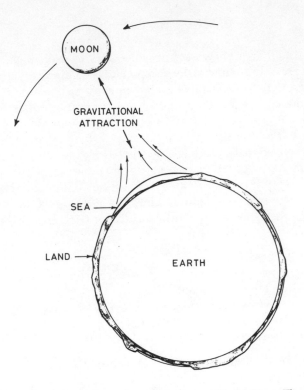

FIGURE 13.9. The crust of land on the earth's surface has variable thickness. The gravitational force of the earth holds the sea on the crust's surface. When the moon passes close to the sea, the sea is attracted by the moon's gravitational field as well, causing that part of the sea close to the moon to rise. This is the origin of tides, which could be used as a source of renewable energy. The distance between the moon and earth is 30 times the earth's diameter.

at high tide and release it at low tide, the water will fall down to the lower level, creating an artificially induced waterfall. If we force the falling water through ducts and put some rotors in these ducts, the rotors will turn, and can drive electric generators. Electricity is then produced—tidal electricity (cf. hydroelectric power). Further, if we cause the electricity to electrolyze water, we can produce hydrogen fuel from the tides, suitable for storage and transport of the generated energy.

Let us suppose we have a tide of 10 m. This would be an unusually high tide, but there are tides of this height in several parts of the world (e.g., in northwestern Australia near Broome, and in the Bay of Fundy in Nova Scotia, Canada). Suppose, further, that we allow the water to flow into an area 5 km along the coast and 2 km back into the land. Let us suppose that the water flows out again in 12 hr. It can be calculated (see Figure 13.10)

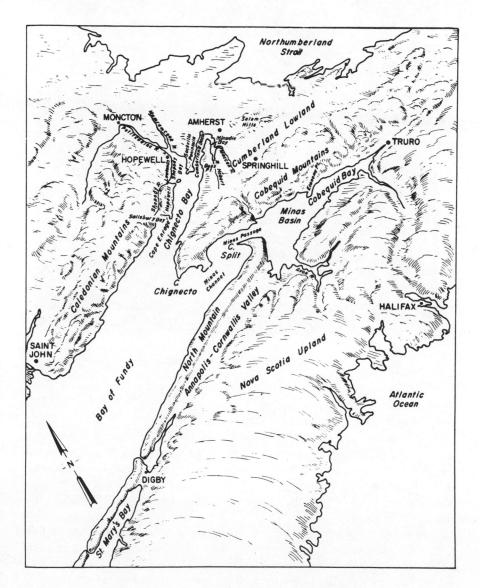

FIGURE 13.10. Topography of the Bay of Fundy region.

FIGURE 13.11. The tidal energy generator of Rance.

that we would have an energy source of some 250 MW.

In Normandy, France, near Rance, there is a tidal energy conversion (Figure 13.11) plant which has been operating for several years. The electricity from the plant feeds into the French electricity system. However, one difficulty of looking for energy from the tides is that there are not enough places in the world where the tide is sufficiently high to make building the dams worthwhile.

We have to search around the world to see where we can find places with tides of 10m or more. No reports of such a study has yet been published. However, it does not seem likely that tides will give us a major portion of our energy, although it may be worthwhile building the energy conversion machinery at some places, such as in northwestern Australia or at the Bay of Fundy in Nova Scotia. Figure 13.12 shows a tidal generator at Rance.

Energy Beneath Our Feet

It is hard to accept on the basis of everyday experience, yet it is a fact that earth is essentially a container full of liquids. The liquid inside is hot (900–1000 °C) and consists of molten rock which contains a variety of metals.

We know something about the constitution of this liquid because the material which spurts out of volcanoes comes from the liquid part of the earth. When lava falls down upon the surrounding terrain, it solidifies, and a cone-shaped volcano seen so often around the world is the result. In fact, this is one method by which earth's land masses are created. Even today, islands are being formed by volcanic eruptions.

FIGURE 13.12. A rotor from the tidal generator of Rance.

The thickness of the earth's outer rock skin (Figure 13.13) on which we live averages 60 km and the radius of the earth is roughly 6000 km, so only about 1% of the distance from the surface of the earth to the center is rock. The rest is liquid. Some geologists think the core of the earth may consist of molten iron.

Near the surface of the earth the temperature is about 600 °C and the temperature inside is around 1500ºC. We have a 900° temperature difference which we could use for driving a heat engine. (In the ocean thermal gradient method of converting solar energy to electricity—see Chapter 10—the temperature difference is only 20 °C, but of course it is easier to flow warm 25 °C water through a boiler than molten lava at the very high temperature of 900 °C.)

Upon first consideration, the idea of boring a shaft through the earth's outer, rocky shell down to reach the magma in the earth, and then inserting tubes which contain water that would turn to steam seems preposterously difficult. The deepest holes made so far are 20 km, and the rock in earth's shell is about 60 km thick.

However, we do not need to go completely through to the core to get to rocks at temperatures which would provide useful amounts of heat. At about 5 km deep the temperature is about 300 °C, which is more than enough to turn water to steam. Thus, there is a chance for obtaining useful, satisfactory energy extraction (see below) at that more practical depth. Geo-

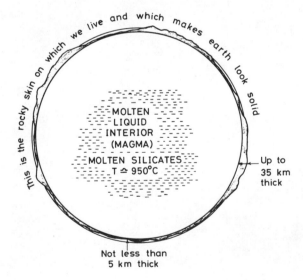

FIGURE 13.13. The planet earth as a sphere of molten silicate encased by a thin layer of rock.

FIGURE 13.14. Dr. S. H. Wilson had much to do with the generation of geothermal electricity in New Zealand.

thermal energy in New Zealand was explored by Dr. S. H. Wilson (Figure 13.14).

How Hot Rock Geothermal Energy Might Become Practical

To get large amounts of energy from the heat inside the earth, a cavity would be blown in the rock at a depth of a few kilometers, as shown in Figure 13.15. Cold water flows down one tube, and steam flows up the other. A temperature of 300 °C is more than enough to form steam from any water injected into the rock, and, therefore, hot rock geothermal energy would seem at first to provide a virtually infinite source of heat energy for us. A schematic of a geothermal reservoir is seen in Figure 13.16.

On the surface, when the heat comes out, it would be used to work turbines and produce electricity. Some of this could be made into hydrogen by electrolysis for use in transportation and industry.

Difficulties in the Attainment of Hot-Rock Geothermal Energy

There are many difficulties in attaining hot-rock geothermal energy in a practical sense. The major problems can be summarized as follows:

1. *How would we form the cavity?* It is clear that we could form small cavities in the earth. This can be done with conventional bombs, exploded under the ground, and it might be that we could use small atomic bombs,

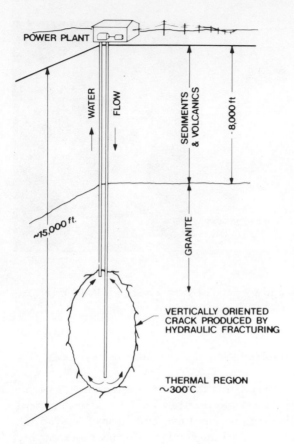

POWER PLANT

WATER FLOW

SEDIMENTS & VOLCANICS

~8,000 ft

~15,000 ft.

GRANITE

VERTICALLY ORIENTED
CRACK PRODUCED BY
HYDRAULIC FRACTURING

THERMAL REGION
~300°C

FIGURE 13.15. A dry-rock geothermal energy system developed by
hydraulic fracturing.

exploded deep within the rock, to make sizeable cavities. However, it seems impractical when we calculate how big the cavity must be to supply a small town; it turns out that it would have to be several kilometers in diameter. On the other hand, one might think of making many smaller holes, each about 100 m in diameter. This is more feasible.

2. *What about eventual cooling of the rock?* As we bring water to the hole, we are bringing new *cold* water into contact with the rock's *hot* surface. Eventually, the hot surface begins to cool. After some years, the surface will be below 100°C and no steam will be formed when more cold water is injected from the surface. We would have to abandon the hole and wait some 10 years for it to heat up again before we could re-use it as a source of steam. (Of course, in the meantime, we could be using other holes.)

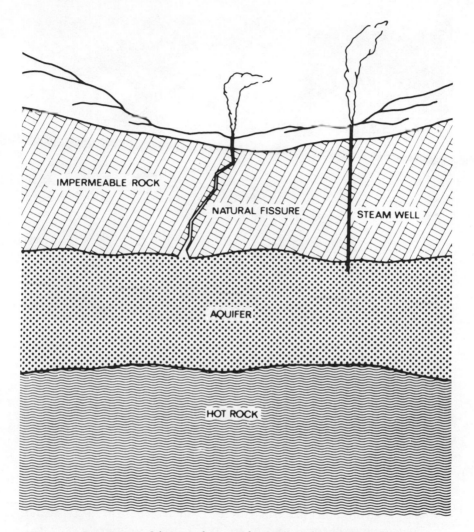

FIGURE 13.16. Schematic diagram of a "wet" geothermal reservoir.

3. *Pollutants.* As the hot water acts upon the sides of the rock, it splits small quantities of the rock off and this will come out with the steam as pebbles and gravel. Many tons of material per day would come out with the steam, and would have to be removed before the steam reached the turbines, for if it stuck in the turbine blades, these would become pock-marked and eventually break.

Figure 13.17. Part of the bore at Wairachi in New Zealand shows the various components making up a well and its equipment—separators, silencers, branch lines, etc.

Low-Grade Geothermal Energy

This type of geothermal energy refers to the hot springs which are found at many places around the earth. Here, steam spurts forth from below the earth's surface. This kind of "low-grade" (temperatures below 150°C) geothermal is used to heat boilers to make electricity. Significant amounts of electricity are produced this way in Southern California, New Zealand, and a few other places. Figure 13.17 shows a bore in New Zealand.

Low-grade geothermal sources are by no means completely exploited. Significant electricity could be extracted from them. However, even if all the low-grade geothermal sources in the U.S. were used, it would mean only a 5% increase in the energy supply for the country, where the energy demand has been increasing for many years at a rate of 7% per year.

Summary of the Prospects of Geothermal Energy

The attractive aspect of the geothermal energy concept is that the energy source itself, the hot liquid within the earth, is so large that we could exist on the heat from it for thousands of years, i.e., we could *in principle* go on taking as much heat out as we needed to support a high population on earth, without causing a significant diminution of the temperature. The problems are mainly in the engineering of enough holes, each with a large enough area, to produce practical, inexpensive energy. Again, the difficulties are in the engineering of a concept which is theoretically sound and well understood, unlike the Pandora's Box of nuclear energy.

14

Energy Storage
and Transmission

Energy Carriers: A Choice Among Three

In the lifetimes of the present younger generation, raw energy will have to be produced not from oil or coal, but from solar or atomic sources. Our reliance upon fossil fuels will end within the lifetime of many of the readers of this book, and we must build a system for collecting and distributing inexhaustible clean energy in less than 50 years. If we fail to do this, our civilization will be in doubt.

But, so far, we have dealt only with one side of the coin. We have shown how it is possible to obtain energy from nuclear reactions (Chapters 6 and 8), pointed out dangers associated with some reactors (Chapter 7), and the promise of others (fusion reactors, Chapter 8) which might be available later in the next century. We have fastened upon solar and wind energies as those which must be developed and used in the immediate future, and which are clean and abundant, though dilute, and collectable only in large amounts over a large number of square kilometers.

When we obtain solar energy, as explained in Chapters 9 and 10, it can only be in the form of electricity or heat. We could use it in the form of heat, or convert it into electricity by exposing photovoltaic cells to sunlight.

Heat is not a convenient form in which to *store* energy. We cannot, for example, put it in a car. Electricity is not a good form in which to send energy over very long distances, for at distances over 500 km, there is too much energy lost in overcoming the resistance of the metallic-conductor transporters. The energy at the receiving end is hence unacceptably expensive, because much has been lost on the way. What we need, therefore, is not only energy *sources* (atomic, solar), but also energy-storage *media*, meaning either forms of energy, or substances which can store energy (potential energy), which can release such energy on demand.

What characteristics must these media have?

1. They must be easy to transport.
2. They must be easy to store.
3. They must be convenient to use—their potential energy must be easily converted to usable energy, such as heat, electricity, mechanical energy, and light.

SOURCES AND MEDIA

The difference between sources and media is a very important point. Energy sources are typified by solar energy. The sun is the basic source of energy in our solar system, but it does not store energy. It is *creating* energy all the time and radiating it to us.

Electricity is a medium of energy. There is no place that is a natural source of electricity, where we can dig it up, find it, or collect it. Electricity is "messenger energy." It transforms energy produced by natural sources, such as fossil fuels, or the sun's rays, or winds and tides, and travels to user places in a convenient way.

WHAT ARE THE POSSIBLE MEDIA (CARRIERS OF ENERGY)?

Electricity has already been mentioned as a carrier of energy. It is a convenient, economical media for transport over short distances, i.e., less than 500 km or so. For use over longer distances, however, transport via electricity becomes inefficient and uneconomical. When thinking of transferring energy over intercontinental distances, from places where solar energy is collected to places where it is most needed, transporting energy over some 5000 km or so will often be necessary. In these cases, transmission of electricity through wires would be too expensive.

Another possible energy medium for transporting energy is hydrogen. Hydrogen has several attractive aspects. It can be sent over long distances in pipes, and requires only small amounts of energy to move it along. It can also be liquified and transported in tankers.

However, there are also other advantages of hydrogen. It is easy to store. It can be used in the chemical industrial processes for many large scale reactions, e.g., the making of ammonia, fats, etc. When it undergoes combustion with oxygen, giving out heat energy, it forms only water, which is nonpolluting. The ideal ecological cycle, as pointed out in the introduction to this section of the book, is to take water, break it down with electrical or heat energy, forming hydrogen and oxygen (again, no pollutants are produced), and use the hydrogen as fuel. It is burned in a combustion cycle, or used in a fuel cell, in either case producing water, returning us to the beginning of our cycle. This cycle of decomposing and reforming water can go on forever in a completely harmless way, particularly as far as the environment is concerned.

There is a third medium which is also a candidate for an energy medium, and that is methanol (methyl alcohol). What advantages does methanol have over the simpler alternative of hydrogen fuel? One is that hydrogen is more difficult to handle than liquid methanol. Hydrogen is a gas at room temperatures, and difficult to liquefy. We can store it in cylinders as a gas, or we could liquefy it and use storage tanks to send it around in liquid form. However, these methods give rise to technical difficulties. For instance, the tanks have to be well insulated, and would thus be expensive.

Therefore, for certain purposes, it may be more practical to make methanol from hydrogen, reacting the hydrogen with coal*, and to use the methanol as a fuel. Projected methanol costs are listed in Table 14.1.

Pros and Cons of the Various Media

Of the probable energy media—electricity, hydrogen, and methanol—none is ideal for all purposes. However, each has its realm of utility. Electricity is convenient to use in a household because it is easily transported over short distances, and easily switched on and off. Methanol would be convenient for cars, because it can be poured into tanks and transferred as gasoline now is. Hydrogen would be ideal for massive use in industry, because it

*A couple of generations from now, in the post-2050 era, when much of our coal may also be gone, carbon dioxide could be extracted from the atmosphere and combined with hydrogen to make the methanol.

TABLE 14.1. PROJECTED COSTS OF LARGE-SCALE METHANOL
PRODUCTION (COST PER M BTU)

Year	Coal gasification to methanol	Anaerobic digestion of vegetable matter	Nuclear electrolysis to hydrogen, CO_2 + hydrogen, etc.
1973	$2.25[a]	$3.00	$4.50[b]
1974	$4.50[c]	$3.45	$5.17[d]

[a]Coal at $5.00 per ton.
[b]1973 dollars, 10% return on investment.
[c]Coal at $10 per ton.
[d]1974 dollars, 15% return on investment.

gives off no pollution. In aircraft, hydrogen would be good because it is light in weight (for a given amount of energy).

Each of these energy media also has negative aspects. Electricity is expensive to send over long distances. If factories were run entirely on electricity, we should need to develop more electrochemical processes, and much more research in electrochemical engineering would have to be done. This would be difficult because in the recent past, we have not trained many electrochemical engineers in the U.S. We can get away with less research, less training of engineers, and less time, if we use hydrogen and continue the use of well-known *chemical* (rather than the less well-known electrochemical) processes.

The negative side of hydrogen, as mentioned above, lies in its handling. Gases are more difficult to contain than liquids. Hydrogen must not be allowed to leak out into an unventilated, closed space because, for a large range of ratios of hydrogen to oxygen, mixtures of the two are explosive if ignited. Methanol eliminates the difficulties of handling hydrogen and avoids the danger of explosion. But it has troubles of its own. For example, methanol exhausts in cars would be as polluting as present gasoline exhausts.

TRANSMITTING ENERGY OVER LONG DISTANCES

Hydrogen could be distributed over long distances more cheaply than electricity

In looking at the possibilities of getting clean abundant energy from inexhaustible sources, rather than from fossil fuels, we have warned of the dangers of the coming exhaustion of fossil fuels, of the rising prices of these, and of the need to accelerate, and in some cases redirect, research and development. We have stressed the *possibilities* which science and engineering offer, without much mention of the cost of this technology. We must, at this point, ask whether these possibilities are practical, making the projects economically profitable, so that some entrepreneurs (and this includes governments as well as private organizations) would want to do it. It must be clearly realized that there is only interest in having a method for producing energy if it is at a price *which people can afford.* From the viewpoint of the average citizen, indeed, this is the central point of the energy picture. We will buy the form of energy which has the lowest cost, paying only as much attention to questions of cleanliness and safety as we can afford.

One of the major contributions to energy prices for the consumer is the shipping cost of sending the fuel to the place where it is needed. The cost of transporting the energy may be half the total cost for the consumer.

When we look at the proposed new sources of energy, this situation of transportation of the fuel from producer to user site comes into focus. The sources of abundant clean energy are principally solar. Since the areas where the sun shines most are far distant from areas where the energy is needed, the method of getting energy from one place to another is economically important. So, how do we get the energy from there to here at reasonable prices? Derek Gregory (Figure 14.1) showed hydrogen to be cheaper to transport than electricity.

WHY LONG DISTANCE ELECTRICAL TRANSMISSION IS NOT ACCEPTABLE

We could do various things to reduce the loss of electrical energy which occurs when we send electricity down a cable. We could use a larger diameter wire, which would offer less resistance to the current. However, the extra wire needed would be costly, and weigh more than the wire with the smaller cross section. The resulting increase in the price of the installation would more than balance the decrease in energy price arising from the lessened loss of energy due to resistance, increasing rather than decreasing the cost of electrical transport.

Over short distances, up about some 500 km, electrical wire systems can be economical and convenient. In general, there will be a grey area between about 400 and 800 km where using a wire grid to send energy in its electrical form may or may not be the economic method of choice. Below 400 km, it will be economical, and above 800, prohibitively expensive.

FIGURE 14.1. Derek Gregory was the first, along with Ng and Long, to publish calculations showing that it is cheaper to send energy in the form of hydrogen than in the form of electricity, if the distance is sufficiently large.

Hydrogen Could Help Reduce the Cost of Sending Energy Over Long Distances

The distances of interest for energy transport in the rest of the discussion in this chapter will be those greater than 500 km. Suppose, as seems likely, that an important component of energy after the year 2000 will be solar energy. The main areas of the world where the solar energy can be obtained in abundance are south of the 30th latitude north and north of the 30th latitude south, i.e., some 3000 km on either side of the equator. To go to London, Tokyo, or New York from these latitudes will be a few thousand kilometers, and therefore it is out of the question to use an electric grid for the transmission of energy. Using a grid, the energy at the point of reception would be too expensive.

However, suppose that we made hydrogen from solar energy right where it is collected, in the equatorial regions. Then, as we explained earlier in Chapter 12, we could electrolyze water *locally* and send the hydrogen that is produced through pipes over long distances, with little increase in cost. It takes little energy to push hydrogen through pipes because hydrogen gas has a low viscosity, and it is the viscosity of a gas which determines the ease of pushing it.

When the hydrogen has to cross water, a different economics may apply because running the hydrogen-containing pipes in and out of the water may be difficult and expensive. At sufficiently large distances, it may be cheaper to liquefy the hydrogen gas and take it in tankers, as liquefied natural gas is transported today.

For distances under 500 km, it is no longer any cheaper to send energy as a gas through pipes than electricity through wires. Although it is always cheaper once the energy is in a gaseous form, it costs money to get it in that form, and is therefore not the method of choice for short-range transport.

Very Long-Distance Transmission of Energy

We have been talking about the transmission of energy over long distances, around 5000 km. When the distances become even greater, there is a different method proposed for sending energy which calculations show to be cheaper than the method of sending hydrogen through pipes. This idea is to send energy by means of electrically produced microwave beams through space, to satellites, which then reflect the beams down to earth again, to energy-consuming cities and developments (Figure 14.2).

Even with beamed power, one would need hydrogen as a medium of energy in the end. The production of energy from fusion (Chapter 8) or

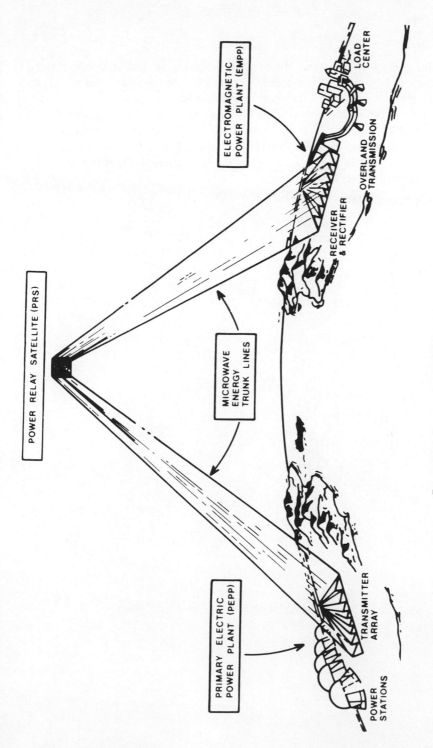

FIGURE 14.2. Power relay satellite concept.

solar energy or the other alternatives discussed would be continuous, and the cycle of use time in the different parts of the world is *not* continuous. One still would need to *store* the beamed energy in some form upon arrival, and the cheapest massive storage is likely to be to convert the energy first to electricity, and then to hydrogen (as seen earlier in this chapter).

Satellite transmission is truly a space-age concept. We can expect to hear more of it as the ideas of long-distance transmission of our new energy sources are proposed.

Part III

The Hydrogen Economy

A concept will be presented here which came into prominence in 1973, from a suggestion made in 1971. Hydrogen gas will probably be the most readily available fuel in post-2000 times. It will first be produced by electrolysis of water, using electricity from hydroelectric sources, and the cost should be lower than that of gasoline in the mid-1980s.

The first priority for development of the needed, new supply of energy is deciding finally on which energy source is to be used—atomic, solar, gravitational, etc. A second task is to avoid polluting the atmosphere as a result of burning fuel. For example, when we burn gasoline, useful heat energy is produced, but the side effects are the production of long, complex, "unsaturated" hydrocarbons which fix themselves onto dust particles and form smarting, obscuring smog which endangers our health and uglifies our environment.

We need to do away with pollution from burning hydrocarbon fuels for energy. In the future, we must look to the inexhaustible solar source, and learn to run industry and transportation systems in a clean way: It would be attractive to run *everything* on hydrogen, a universal, clean fuel. That is the idea of a hydrogen economy. Figure III.1 shows Professor Justi, a proponent of the use of hydrogen as fuel.

The hydrogen economy may be defined as a situation in which the majority of industrial, transportation, and household uses of energy are provided by hydrogen. It is the all-hydrogen view of the energy future. Households would receive hydrogen through pipes. This hydrogen would be used for heating by direct burning, and also be used to produce electricity through fuel cells (Chapter 12) in the basement or on the roof of the building. The heat left over from the operation of the fuel cells to produce electricity could be used to heat the buildings concerned (at central plants such heat is now wasted).

For transportation, hydrogen would be used to run cars. It would be

FIGURE III.2. Professor E. Justi was one of the
pioneers of the idea of hydrogen as a universal fuel.

stored in gas cylinders or in liquid form or in a hydride, and used to work
internal combustion engines in the cars, or else be utilized in the fuel
cell–electric motor combination.

In industry hydrogen would replace natural gas or oil to power
machines and industrial devices thus eliminating air pollution from the burning of the fuel. The only product with hydrogen would be pure water
vapor, some of which could be collected and used as drinking water.

Another way of running industrial equipment and transportation in the
hydrogen economy is to use the following scheme:

$$\text{Hydrogen} \rightarrow \text{Fuel Cell} \rightarrow \text{Electricity} \rightarrow \text{Electric Motors}$$

There are two separate advantages of such a scheme, apart from the
removal of air pollution by the use of hydrogen. One advantage lies in the
greater efficiency with which a fuel cell converts hydrogen to mechanical
energy, compared with the 25% efficiency of most combustion engines
(which get mechanical energy from burning fuels). In a fuel cell, electricity
would be produced by hydrogen at about 60% efficiency (under some circumstances higher values can be obtained). The mechanical energy comes
from this electricity in a motor at about 90% efficiency, so that the overall
efficiency of the electric-motor-fuel-cell combination is more than 50%.

$$\text{Hydrogen} \xrightarrow{60\%} \text{Fuel cell} \rightarrow \text{Electricity} \xrightarrow{90\%} \text{Electric motor} = 54\%$$

Another advantage of the fuel cell approach is that electric motors have
a number of mechanical advantages over internal combustion engines.

They are much easier to control. The torque they produce is less dependent on having a high speed of rotation than is that of an internal combustion engine. But the nicest advantage is in lessened maintenance costs—an electric motor will go a million miles without needing overhaul.* An internal combustion engine needs overhaul every few tens of thousands of miles.

The hydrogen economy offers practical advantages. More than that, it is philosophically attractive. It is the completely satisfactory ecological scheme for the distribution of energy. It does not upset the balance of nature in any way. It decomposes water to hydrogen and oxygen, which then recombine to form water. The only "net" product of the reaction would be free, usable energy. No pollutants would be produced. This cycle is summarized in Figure III.2.

Thus, in a hydrogen economy, no harmful waste products remain to enter the atmosphere, or distrub the balance of nature. One starts with water and ends with water. Inasmuch as we can collect this energy in "potential" form, we can convert it into forms which would operate our civilization without any damage or unbalancing of nature. Why should we not be able to proceed indefinitely with the sun as our clean energy source,

*Today, the electric motors in diesel electric railway engines are over-hauled only once every million miles.

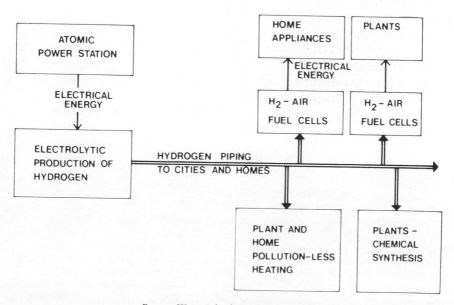

FIGURE III.2. A hydrogen economy.

affording all human beings the opportunity to live, work, and achieve in comfort?

> *One of the eventual results of a wide-spread hydrogen economy would be super-national authorities. The formation of a world-government would be hastened. The danger of disastrous radioactive pollution as a result of a conflict involving the use of atomic weapons would be reduced.*

Another advance in civilization could be attained in the development of a solar-hydrogen economy on a large scale. The main energy-producing centers of the world could each be very large in area, and perhaps restricted to about 100 in number. The price of the energy would decrease with the increase of the size each area. The world would be covered with pipeline and tanker networks distributing hydrogen. As living standards depend on the availability, and hence the price, of energy, average living standards would increase. A considerable ecological advantage could be achieved. But another advantage would be likely to follow. "Nations" could be so connected together with energy streams in the pipeline network that no interruption of this flow would be feasible. The analogy of arteries to the body is appropriate. A world government would be a necessary development. Its achievement seems an obvious part of geopolitical evolution of mankind.*

*Yes, we are being utopian. Shouldn't this, then, have been the case with fossil fuels, which are also localized in a relatively small number of locations? Isn't it more likely that powers would vie *against* one another for control of the solar energy plants? There is a major difference between these two situations. Solar energy is essentially inexhaustible, and the sun is not easily owned or fought over, as are fossil fuel deposits. Only the collection and transport apparatus could actually be controlled privately, or by one particular government. In terms of cost, the economy of an inexhaustible energy supply and media is much less prone to disaster than that of an exhaustible supply, but does not offer anywhere near the same profits to the controllers and distributors.

15

Methods of
Mass-Producing Hydrogen

INTRODUCTION

If hydrogen is to be the most important energy medium for transport over long distances, as discussed in the previous chapters, we must find out how to produce it on a massive scale. The cost of the production process we choose will have to be low enough so that the energy from the system will be affordable. Two large-scale methods of preparing hydrogen have been developed. The first is called the *cylinder thermal method* and the second is called the *electrochemical method*.

It is obvious that water is the only substance we can regard as abundant enough with respect to getting hydrogen on a very big scale. We do have a great deal of water, including seawater, and it can be decomposed into hydrogen. The whole point is, however, at what cost?

Our present sources of hydrogen will not be satisfactory in the future. We now get hydrogen by decomposing natural gas, but we should not continue to look at natural gas as a long-range source of hydrogen because its price will have become too high within 25 years, which is all the time we have to research and develop a solar-hydrogen economy.

Coal, too, is a problematic source of hydrogen because of the difficulties involved in the use of fossil fuels (pollution of the atmosphere due to the sulfur content in coal) and the difficulty in mining satisfactory quantities in time.

THE CYCLICAL CHEMICAL METHOD FOR PRODUCING HYDROGEN

To decompose water into the needed hydrogen on a big scale, one of the methods we might think of using is simply to heat it. Heat at a sufficiently high temperature decomposes water, and if we separated the hydrogen from the oxygen quickly enough, their recombination back to water in the cooling process could be avoided. However, upon closer examination, this method is not very promising. We would have to heat water to very high temperatures, over 31000°C, and vessels which could withstand these temperatures without melting or cracking are unavailable. So we must abandon the idea of utilizing the direct decomposition of water by heat.

There is no need, however, to abandon the idea of using heat to decompose water. We need only to change our approach from direct to indirect, and we may be able to get hydrogen with a much lower temperature. For example, a metal will react with water if heated. One of the products would be hydrogen, the other would be a metal oxide, i.e., a molecule containing the metal and oxygen. Then, upon further heating, the metal oxide decomposes to form the metal and oxygen. Thus, in carrying out these two reactions, we have undergone a cycle—we start with a metal and come back to it again. Such an example is oversimplified, but it illustrates the ideas of a cylical method for producing hydrogen. This example of a low-temperature cycle is shown in Figure 15.1.

The advantage of using the cyclical path, rather than using heat directly on water, is that the temperatures needed are lower. The highest temperature needed in the cycle chosen in Figure 15.1 is 800 °C. This is a high temperature, but attainable. Vessels can be found which can withstand this temperature. Steel, which is a cheap material, could stand the temperature if suitably protected from chemical attack, and we could therefore contemplate cycles of this kind as a possibility for sufficiently cheap decomposition of water.

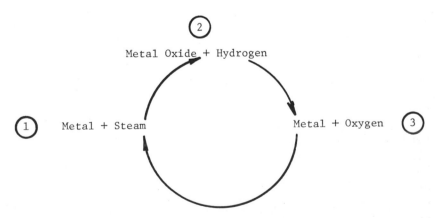

FIGURE 15.1. The cyclical method of producing hydrogen.

We have to apply huge amounts of heat to produce hydrogen on the large scale needed. Therefore, we need to know the highest temperature which we could expect from a large heat source 25 years from now. *Atomic* furnaces yield a temperature of about 900 °C, and much planning has been based on this. In respect to solar energy, virtually any temperature (e.g., 3000 °C) can be attained by using lenses and focusing mirrors to concentrate the solar radiation on a small area. However, there will only be a small area of such high temperatures and a larger area heated from a solar source would maximize only a little above 200 °C, unless we could design suitable, large focusing arrangements.

DISADVANTAGES OF THE CYCLICAL THERMAL METHOD

There are always pros and cons to be considered when choosing a chemical process for use in industry. We have stated a major advantage for the cyclical thermal method. It produces hydrogen at a temperature low enough to use atomic furnaces. Therefore, electricity is not needed. For, per unit of energy, it is more expensive than the heat from which it came, because it is a *derived* form of energy, and energy is lost in making it.

There are also disadvantages of the cyclical thermal method. One is the efficiency of the processes, in other words, the heat produced from the combustion of the hydrogen relative to the heat used to produce the hydrogen. There are many energy losses in cyclical thermal methods. During the cycle, gases have to be pumped, heated, compressed, and expanded. Each of these steps requires an input of energy. So, not all of the heat input goes to producing the hydrogen (splitting up the water), but also into these other purposes. Taking these losses into account, many of these cyclical thermal methods for producing hydrogen are inefficient, some having efficiencies as low as 18 %.

Another disadvantage arises from the effect of heat on the vessels. It is true that 800 °C is an accessible temperature; however, it is still quite high, and materials of the vessels corrode and decay at this temperature faster than they would at lower temperatures like room temperature (25 °C) or even 100 °C.

One final, more subtle point, is the matter of *degree* of cyclicity. We have said that the processes are 100 % cyclical—they should reproduce 100 % of the substances which were put in, except for water, which should be completely decomposed to hydrogen and oxygen. For example, tin bromide should be there at the beginning of a cycle, and be reproduced again with 100 % efficiency at the end. But suppose the reactions do not reproduce each substance in the cycle at 100 % efficiency. Nothing is perfect, and the processes may only go according to the reactions to the extent of, say, 99.8%. In this case, there will be a gradual build-up of the intermediate material (not completely recycled). When we have a plant producing hundreds of tons of hydrogen per hour, the solid material piling up from the unrecycled material, 0.2% loss can be huge.

So, like everything else, there are both good and bad points about the cyclical chemical methods, and at this time, it is not yet clear whether they will be practical for production of hydrogen on a massive scale. It depends on the price of its final product, i.e., the price of hydrogen per unit of energy it would produce when burned with oxygen. What price a unit of hydrogen will be by this method, compared with the price of hydrogen produced with some other method, can only be answered after a few more years of research have been done.

The Electrochemical Method of Obtaining Hydrogen

This is a familiar method, already mentioned frequently and described in the previous chapters. It is the electrolysis of water to form hydrogen and oxygen by means of an electrolyzer, and it can be done on a big scale. The type of vessel which has been used in small scale experiments is shown in Figure 12.11.

There is a negative point about this method which warrants explanation. Referring back to the ideas which we have given about sources of energy and energy media (Chapters 13 and 14), it is clear that electricity is needed for electrochemical hydrogen production, and electricity is a *medium* of energy, not a source of energy. We have to pay to get electricity (i.e., have some loss of initial energy) before we get hydrogen via electricity. Now, making electricity is fairly inefficient, at best 40% efficient, and in most cases more like 35–38%. Rounding this figure off and assuming a general one-third efficiency (i.e., 33%), this means that we lose up to two-thirds of our initial energy input (for example, heat from coal) which is being converted to electricity. Thus, as far as hydrogen is concerned, we *start* with a two-thirds loss situation in the electrolytic method, independent of the efficiency of the method for hydrogen production itself. Indeed, this latter figure is very high, at least 85%, but one-third of this is still very low, about 28%.

In the cyclical chemical method, we apply heat energy *directly*. From this point of view, the cyclical chemical method might be better, since we avoid energy loss during the conversion from heat to electricity. However, as we have seen above, we also lose energy in the cyclical method because of the pressure and temperature changes of the reactants in the cyclical steps.

Since it is likely that the electrochemical method will be adopted for massive production of hydrogen fuel, it is important to understand the unique characteristics of electrolysis:

1. It is a simple method, and has been carried out all over the world.

2. The hydrogen and oxygen produced are *separated* by this method, i.e., are given off at different electrodes (see Figure 15.2), so there is no problem getting pure hydrogen.

3. The method gives *exceedingly* pure hydrogen, in addition to being separated from oxygen, it also contains no chlorine, sulfur dioxide, carbon monoxide, or other impurities which tend to be in hydrogen obtained by other methods, like gasification of coal.

4. There is no problem of build up of undesirable side products as in the cyclical methods.

5. There is no problem of rapid wearing out of vessels due to high temperatures. Maintenance would be cheap.

These are powerful points economically, but modern research is needed to produce large-scale water electrolyzers which incorporate the results of recent research concerning the electrical evolution of hydrogen and oxygen.

A modern version of an electrolyzer is shown in Figure 15.3. During the last few years, research has succeeded in lowering the voltage required in the production of hydrogen, and hence the costs of large-scale electrolytic production would be lowered if some of these newer, small-scale experiments were scaled up to industrial level.

Figure 15.4 shows the performance of different designs of electrolyzers. Many people (Bacon in England and Justi in Germany) as well as firms (Al-

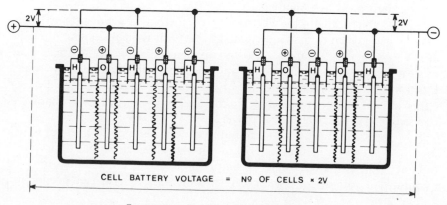

CELL BATTERY VOLTAGE = N⁰ OF CELLS × 2V

FIGURE 15.2. Unipolar cell construction.

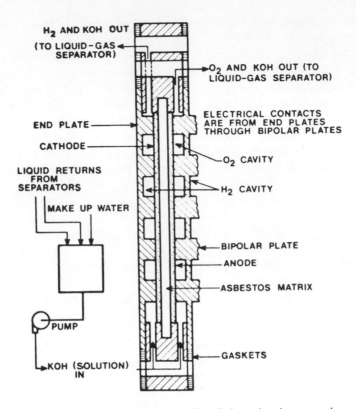

FIGURE 15.3. Schematic design of end cell in Allis-Chalmers bipolar water-electrolysis cell.

lis-Chalmers and Westinghouse in the U.S.) have been busy trying to make
the hydrogen electrolysis process work at lower voltage which would give
cheaper hydrogen (see Figure 15.3).

Thus, the future for the electrochemical method of producing hydrogen
may be better than we had anticipated, so long as we can decrease the voltage requirements. Perhaps we will end up using solar energy to create heat,
use the heat to create electricity, and the electricity to create hydrogen.
Pipes will send the hydrogen to far-off places, and thereafter it will be used
either in heat engines to produce mechanical power or, more probably, in
fuel cells, to produce electrical power.

We probably need about ten more years of research to find out which
method will be best—the cyclical or the electrochemical. At this time we can
only say that the latter would be simpler and easier, but research must be
continued in both before a decision regarding massive hydrogen production
can be made.

Getting the Energy Back from Gaseous Hydrogen at the User Terminal

Using solar-generated hydrogen, solar collection might be in Arizona, say, with the user terminal in Boston. At the point of arrival of the hydrogen, we may choose to convert it into electrical energy again.

Let us imagine distribution centers of the future, where hydrogen arrives from a distant point. The hydrogen will arrive in long-distance pipelines, and distribution will be made to various subcenters—some for industrial use, some for homes, and some for transportation.

We have had good experience with piping of natural gas into households, for use in gas cookers and heaters. It will not be more difficult to send hydrogen through pipes, although they should be different pipes (e.g., their joints more carefully made) since hydrogen escapes more easily through small pinholes and cracks than does methane. Furthermore, hydrogen contains less energy than methane per unit volume, so *more* hydrogen per unit time would have to be pumped than methane to get the same energy. Hence, the tubes would have to be wider, or the pressure greater (and hence the pumping devices bigger), than for piping the same energy quantity of natural gas.

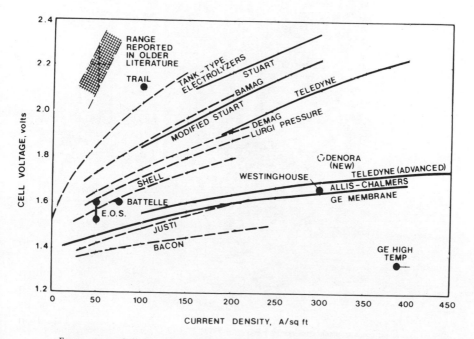

FIGURE 15.4. Cell operating performance of various advanced electrolyzers.

For transportation, hydrogen must be pumped to the filling stations (at last appropriately called gas stations). If hydrogen *is* used in the future for transportation (rather than its rival, methanol), we shall have to pump hydrogen to the filling stations so that cars can pull up and obtain their hydrogen gas pumped into cylinders in place of the gasoline we now use. Likewise, hydrogen would be pumped through pipes to industrial parks, where it would be used as a fuel in the same way as natural gas.

What of Homes and the Electricity We Now Use in Them?

There would be two methods to be used here. In one, we could use hydrogen from distant places in local electricity-generating stations, in or

Figure 15.5. Full-scale stack module for experimental 1-MW fuel cell pilot power plant comprises 376 individual cells. Cell stacks are connected in parallel in power plant. (Courtesy of Pratt and Whitney.)

FIGURE 15.6. From natural gas to electricity for the town's supply; a simulation of a fuel cell module being added to an already existing stack. Natural gas enters the cells and electricity is made in them silently and without moving parts, by electrochemical means. (Courtesy of Pratt and Whitney Aircraft.)

around towns as are now used, and continue to distribute electricity as before. At generating stations, hydrogen would be used as fuel in the turbines which drive the generators which make the electricity. Subsequently, it would be pushed through wires over a few kilometers in towns. This might be preferable to building new pipelines under the city streets.

Another method would be to build fuel cells (as shown in Figures 15.5 and 15.6) to use for producing electricity. Where would these fuel cells be? One possibility is large central power stations. Then, householders would get electricity through wires a few kilometers in length, as they do now. On the other hand, as hydrogen would have to be piped to houses for heating, it might be better to have a small fuel cell in each house to produce the electricity needed for lighting and for electric devices, and to use the heat left-over for heating the house.

THE ADVANTAGES OF USING FUEL CELLS

There is one aspect worth stressing at this point, and that is the *greater efficiency* fuel cells have in producing electrical energy from hydrogen, compared with the efficiency obtained by driving turbines which would then drive electric generators. To get an idea of the efficiency, consider the great simplicity of the fuel cell. You put the hydrogen in and get electricity out, directly; no moving parts.

The main point is that a fuel cell will give electricity with an efficiency of more than 50%, even up to 75%, whereas with the indirect method (hydrogen to turbines, turbines to electrical generators, and so forth) the efficiency is about 35–40%. So, we would get at least 15%, perhaps up to 35%, more energy if we used fuel cells to produce electricity, rather than older, indirect methods of turbine to generator. This is a powerful argument for fuel cells, and we hope to see more use of them in the future.

16

The Storage of Abundant Clean Energy

INTRODUCTION

People talk about solar energy as though it were new. But we have been using stored solar energy ever since we started to burn wood, a product of photosynthesis.

The storage of energy is not of much concern in our present fossil-fuel energy system. But remember that energy is *stored* in the fuels we use: coal, oil, and natural gas. As we have pointed out before, energy of the fossil fuels is *stored solar energy.*

Whenever we see a coal mine, an oil field, or surface equipment which indicates that natural gas is underneath the ground, we are observing the presence of energy stored hundreds of millions of years ago. We leave the oil, coal, and gas in the ground, stored there until it is needed.

The artificial stores of energy we keep above the ground at the moment are small. We do have oil tanks in parts of the world,* and we sometimes store coal near electicity-generating plants. But these are only tiny sub-stores of the fossil fuels we still have in coal mines and oil fields.

In the new energy age which we are approaching, there will be problems with respect to storage of energy, whether it be atomic, solar, wind, or some other source. One reason for the difference is apparent when considering solar energy—it will not be satisfactory to feed energy directly from

---

*The oil storage depot at Russ-El-Tunurra, in Saudi Arabia, and in the Persian Gulf, is the largest in the world, and takes oil directly from the vast fields which underlie the Saudi Arabian deserts. Russian and American satellites pass over several times a day, taking photographs which enable analysts to estimate the amount of oil stored there.

solar cells into the electric grid. Should the weather become cloudy, or after sunset each day, our energy would be gone. So, with solar energy, storage is absolutely necessary in order to have the energy to use at the time we need it.

Methods of Storing Energy

There are many methods of storing energy. We are going to mention six of them, describing a few in detail.

To begin with, we could store solar energy for a few days as *heat*. To do this, we allow the solar heat collected on absorbing solid panels to heat a gas (preferably air), and then we pass the hot gas through a storage liquid which is thus heated. Water could be used as the storage liquid in this case.

A second way to store energy is in the form of *electricity*. We get electricity directly from photovoltaic cells as described in Chapter 10. When we want to store energy in an *electrical form*, we put it into batteries. Of course, these would have to be unreasonably large to accommodate the energy needs of large numbers of people. Therefore, this method is only acceptable when small amounts of energy are involved as, for example, the energy needed to run a car a few hundred kilometers.

Mechanical energy could be stored in a flywheel. A solid wheel has great inertia (resistance to motion), and therefore, when we have revved it up to a substantial speed, it contains within the motion of its high mass a large amount of kinetic energy. We can recover the energy by making the flywheel drive a useful machine.

We can store energy in a *chemical form* by converting a substance containing energy in the bonds between the atoms, such as hydrogen, to a compound like magnesium hydride. We then store the magnesium hydride in the form of a powder. The big advantage over the hydrogen itself is that the hydride is not a gas and can be stored without the necessity of a big, gas-tight container. Hydrogen can also be stored as an alloy of iron and titanium.

Next, we can store hydrogen underground in the form of a gas. This will be one of the main ways in which we shall store *gaseous hydrogen*. The underground spaces could be made artificially, or we could use former natural gas fields, now empty, where methane has already been pumped out and into which we could pump hydrogen.

Lastly, we could store energy in a *liquid* form. An example would be in tanks containing methanol, or perhaps (though it would add 20–30% to the cost due to the cooling requirements) in the form of liquid hydrogen (Figure 16.1), which exists as a gas at room temperature and must be greatly cooled to be liquefied.

FIGURE 16.1. A liquid-hydrogen storage tank at the Kennedy Space Center. The energy contained in one tank is a day's household energy for one-third of a million people. (Courtesy of NASA.)

Deciding which methods will be best has yet to be done, with a stress on cost factors. There will not be only one method, but several, depending on what is to be stored (electricity? hydrogen? methanol?). For example, we might store electricity for cars in storage batteries; methanol in tanks; solar energy for general use could be stored in hydrogen underground or in a liquid made from hydrogen, such as ammonia; wind energy would be converted to electricity, this used to electrolyze seawater to hydrogen, which is then stored in tanks under the sea; and fuel for aircraft as liquid hydrogen.

STORAGE OF ENERGY IN THE FORM OF HEAT

Instead of simply storing heat as a hot liquid in an insulated tank, we can store it by melting solids to form liquids. Potential energy is stored initially by melting a solid. Any solid form contains less energy than its liquid form, and during melting of a solid, this energy is stored in the liquid. Conversely, freezing releases heat. Now, if we pass a gas through the liquid during the freezing process, the gas absorbs the heat. In other words, we are retrieving heat energy from the storage system which we had melted.

One of the liquids which can be used for this process is a mixture of sodium fluoride and magnesium fluoride. It melts at 832°C, and can store 400 calories of heat per gram of the mixture.

Storing Energy in Its Electrical Form

Figure 16.2 shows a nickel–cadmium cell discharging. In Figure 12.4 one of the batteries developed in the past few years is shown. It contains sodium and liquid sulfur, and these react to form sodium sulfide in liquid sulfur.

Both sodium and sulfur are cheap and plentiful. When the battery is charged, i.e., electricity is put into it, sodium and sulfur are formed. Thus, we have stored energy. Conversely, when the battery is discharged, sodium and sulfur re-ionize and return the electrons to form a current, reacting to form sodium sulfide, and the energy previously stored is released to do useful electrical work. For example, the battery can drive the motor of an electric car (Figure 12.1).

The Pros and Cons of Heat and Electrochemical Storage

The main consideration is whether we have energy in the form of heat or electricity to begin with. We can store it in either form. But heat storers will lose energy and cool, so they are useful for keeping energy only for several days at a time. On the other hand, electricity storers can hold energy for a year or two and produce the energy again at the touch of a switch.

There are many other pros and cons one has to think about: weight and cost have to be taken into consideration, and each individual situation has to be evaluated separately. The balance of cost and convenience are the final deciding factors as to which method should be used.

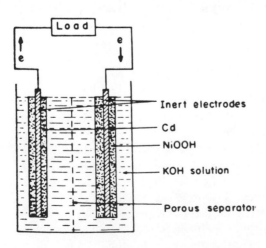

FIGURE 16.2. A nickel–cadmium cell at the beginning of discharging.

Storing Gaseous Energy Underground

Underground storage is an easy concept. It is simply a matter of finding the right storage area. This may be a natural cavity, a cave, or it may be what is called an aquifer, which is a layer of water found underground (it would have to be pumped out, of course, to receive the gas one wishes to store). To make an artificial cavity, one could use explosives (perhaps eventually atomic bombs might be used) to form large cavities underground.

There are other considerations when we get into the details of underground storage. We cannot assume simply that the gas will be stored underground without taking into account possibilities of leakage and the acceptable leakage rate, which obviously has to be small. Every storage situation will have to be subjected to detailed examination and evaluated by engineers on the basis of the most energy stored for the least money and the longest time.

One possibility is storage under water. We have a lot of room under the sea, and we could make use of the pressure caused by the water to help keep stored hydrogen under higher pressure and hence in smaller vessels for the same amount of energy stored.

What Methods Will Be Most Used in Our Time for Energy Storage?

We may have to store electrical energy directly using the photovoltaic method of converting solar energy to electricity. Storing electrical energy suggests batteries. If the batteries have to be so big that they become impractical, as big as a house, say, it may be more sensible to make hydrogen from water by electrolysis and to store the hydrogen underground, where we can arrange to have plenty of room.

Storage of hydrogen energy for houses might be in the form of hydrides, such as iron–titanium hydride, with the hydrogen delivered in pipes and forming the hydride from the iron–titanium. When the hydrogen is needed, it could be recovered by reducing the external hydrogen pressure so that the hydrogen stored in the hydride comes out again (see Figure 16.3).

How about the storage of hydrogen gas to provide energy for an entire city? Storage in the future may often be in hydrogen gas. Natural gas, which we use right now in large amounts for running cities, is often stored underground in large amounts, so this idea is not new.

There is a good deal of concern about the dangers of hydrogen as a fuel because it can cause explosions when ignited with oxygen from the air. Hydrogen in caves under a city would *not* explode, because to get it to ex-

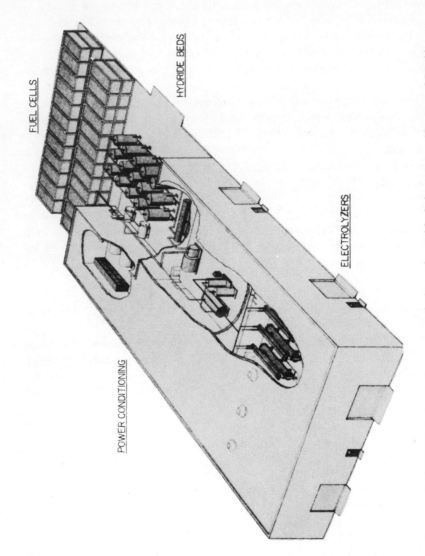

Figure 16.3. A 26-MWe hydrogen energy storage facility with metal hydride reservoirs. (Courtesy of Brookhaven National Laboratory.)

plode with oxygen, you need a confined space, an oxygen source such as air, a flame to ignite it, and a specific range of compositions of oxygen with hydrogen. It is no more dangerous than natural gas reserves, or gasoline pumps in gas stations, or oil wells, except that a broader range of fuel to oxygen proportions are explosive in the case of hydrogen.

For example, suppose we keep hydrogen in large cavities under the ground, letting it out when we want it, and putting it into burners or machines. Danger would come only if a car with a tank containing hydrogen were left in a garage and the tank leaked, and there were no ventilation in the roof of the garage; then hydrogen could accumulate at the top of the roof. Someone must then climb up to the roof and make a spark or flame to initiate the explosion. In the absence of one of these circumstances, no special dangers would arise from using hydrogen as a fuel. To create an explosion, we would need not only a foolish car owner, but one who intended to blow himself up.

Another point should be made about explosions and accidents involving combustion, i.e., fuels such as hydrogen, oil, gasoline, natural gas, etc. These accidents are harmful, destructive, and often deadly. However, they are localized to the region in which they occur. Aside from pollution (in the case of the fossil fuels, not hydrogen), the effects will be known, predictable, and over once the situation is under control. On the other hand, in the case of atomic reactors, we have already pointed out the widespread effects of an accident (Chapter 7). The danger would not be localized, but would spread over broad regions in the form of radioactive pollution, damaging the environment and the health of its inhabitants, not only immediately, but in 10 to 20 years, and even in future generations in the form of mutations (genetic defects).

Beyond
the Hydrogen Economy:
Some Futuristic Ideas

Concepts for the Next Few Hundred Years

One of the greatest possibilities for the future has been described in Chapter 8. It is atomic fusion. There are so many difficulties at present that its practical development may be delayed until after the year 2100. Another concept which is quite possible, though the technology appears long-term, is that of tapping the energy of the magma itself, the inner liquid core of the earth. Will we be able to bore a hole 50 km deep in 100 years? And control what we find? Holes 10 kms deep are already being bored in oil exploration.

Concepts of the Next Few Thousand Years

Obviously, if we think 1,000–10,000 years into the future, we are speculating in vast uncertainties. First, the course of the development of science will have been so tremendous that the concepts of today may be regarded as trivial special cases of far more general laws—or even be proven quite wrong. So, what we write here is only stimulation in thought and more in the realm of science fiction than science fact.* There again, science fiction

---

*A further diminishment in the likelihood of being right is that we may well have encountered advanced civilizations from other solar systems within the time period mentioned here. In fact, it is time scientists stopped attempting to laugh off the UFO phenomenon. Something *is* happening and the simplest explanation (visitors from another planet) should be viewed with a relaxed, logical approach, not in panic.

has an excellent reputation for predicting the science of the next generation (the 1930s fiction had all sorts of space-suited men shining fantastic rays of some kind and signaling to spacecraft from the moon's surface). However, we are looking ahead 50–100 generations in the following discussion, and hence the likelihood we shall hit the mark is small, but perhaps just worth recording.

Among the concepts we should like to mention, one has to do with planetary engineering. Suppose we are, in the future of which we are speaking, easy riders in space. Further, we are very able in atomic geo-engineering and using the force of atomic explosions to split up mountain ranges and rearrange them. We may well be able to do that long before 1000 years. One further step would be to regard the material of this solar system as ours to play with. *We could approach Venus, then, and send down the necessary automated modules. Several hundred are needed and when they get to the surface, they go straight into the planet, each stopping at a certain point within. When all is ready, the command module pilot, standing off Venus by some thousands of miles, activates the telemetric circuits which transmit the message to the atomic explosive devices of the modules inside Venus. The force of a billion Hiroshimas splits the planet into fragments. They continue near-Venus orbits but after a few years, we attack them, and re-explode them into tiny asteroids around 10 km in diameter. These become controllable, and with atomic-powered impulse drive, we have enough energy to impel hundreds of these fragments to a new solar orbit, outside that of the earth. The sun, in fact, is now "surrounded," by, say, 1000 of such Venus fragments. Now, the object of all this explosive far-outness is solar energy trapping. It is the first stage whereby the Third Technological Race could increase its available energy. On each of the Venus fragments will be built solar reflectors, and these will beam the light they intercept from the sun to earth. Over many hundreds of years, our descendants could use the planetary material and atomic energy from hydrogen fusion to build these orbiting collectors. Figure 17.1 shows the concept of the Dyson sphere.

Another concept can only be approached and not described because it is yet too misty. Huge energies are available in the earth's gravitational field, but much larger ones in that of the sun. Will it be possible, 1000 years from now, for us to control a gravity communicator, a very heavy body which orbits earth and sun, and the motion of which could bring to us some of that energy from the sun, energy which already has us in its grip?

Finally, let us refer to a concept even more far out than the above two, and connected with black holes. When stars collapse, they form extremely

---

*Certainly a whimsical and not environmentally sound attitude.

compact bodies. The mass is so enormous that the gravity associated with it interacts with the light it would be giving off and makes the collapsed stars invisible (black holes). Now, let us go past the idea of doing planetary engineering on Venus and suppose we can tow a black hole around. It will in fact eat matter, pull everything toward itself, and collapse it (the matter) to nuclei. But all that flowing matter, dashing to annihilation, will be ionized. And moving charges are currents and could conceivably be converted to energy on a grand scale.

Such fantasies have the attraction of science fiction. But there is a certain danger in tarrying with them because they engender an "everything will be all right then" attitude. At this time, it is very unsure whether technological civilization will survive to a point where such far-out ideas could be considered.

Table 17.1 consists of schemes depicting a variety of ideas for energy collection. Some are whimsical, others relatively practical. They are included to stimulate thought and imagination of those people concerned with our future, and who intend to play a part in molding it.

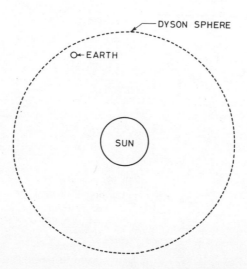

FIGURE 17.1. This figure shows an advanced futuristic idea, which could perhaps be applicable 1000 years from now. At that time, the earth's civilization may be wealthy enough to mount a space operation in which one of the planets (e.g., Venus) is intentionally annihilated by nuclear explosions. The pieces could then be spread out to orbit outside the earth's orbit in a Dyson's sphere, so-named after the originator of the idea. The sphere is meant to surround the sun in three dimensions so that it would intercept the whole radiation from the sun, instead of the fraction which now reaches the earth. In this way, energy could be collected from the sun and beamed to earth.

TABLE 17.1. SCHEMES DEPICTING IDEAS FOR ENERGY COLLECTION

Concept	Description	Status
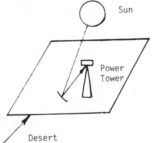	Photovoltaic couple produces electricity. Large desert land areas are needed. Cost of material?	Small test areas only. Awaits capital for mass production to reduce costs.
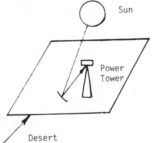	Many mirrors track the sun and concentrate light on top of a power tower.	Test rig operative, Sandia Labs, 1977. A 10-MW plant is being built in Barstow, California.
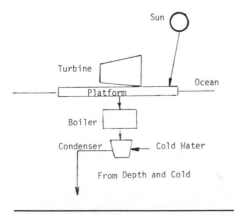	Ocean thermal energy converter. In tropical seas ammonia is condensed with cold water from the ocean depths. The cycle is repeated.	One plant worked in the 1920's. Department of Energy is planning trials in 1980.

Table 17.1 *continued*

Concept	Description	Status
Thermo-Electric Generators — Magma 1000° C, 2000° C — HOT, RUCK	The solid part of the earth is only about 60 km thick, but it is hot. Two different metals in contact produce a potential. At 10 km, the temperature is several hundred degrees Celsius. The thermoelectric devices electrolyze the water which is prevented from boiling. Hydrogen and oxygen separately travel to the surface, giving hydrogen fuel.	Concept only.
Sun — Satellite — Beam of Megacycle Frequency — Earth	The satellite solar collector would be built in outer space by use of the space shuttle.	Plans have been made. Awaits cheaper orbital flights with the space shuttle.
Cold Air — Generator Boiler Condenser — Recycle Stream — Valley, 50-90°C	In this, the cold necessary to condense the ammonia is obtained via many wide pipes which draw air from the cold mountain top. Heating in the valley is always available.	Concept only.

continued

Table 17.1 *continued*

Concept	Description	Status

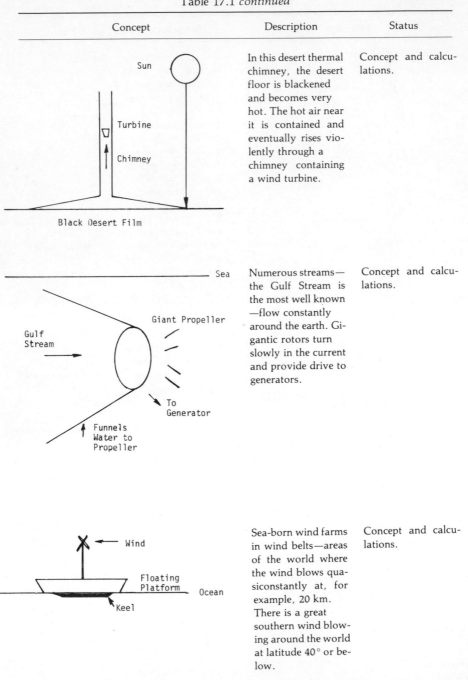

In this desert thermal chimney, the desert floor is blackened and becomes very hot. The hot air near it is contained and eventually rises violently through a chimney containing a wind turbine.

Concept and calculations.

Numerous streams—the Gulf Stream is the most well known—flow constantly around the earth. Gigantic rotors turn slowly in the current and provide drive to generators.

Concept and calculations.

Sea-born wind farms in wind belts—areas of the world where the wind blows quasiconstantly at, for example, 20 km. There is a great southern wind blowing around the world at latitude 40° or below.

Concept and calculations.

Table 17.1 *continued*

Concept	Description	Status
Ice Mountain	There are places in the Antarctic where mean annual winds are over 40 km/hr, i.e., they give eight times more energy per unit area than one would get elsewhere.	Concept.
Floating City Pushes Water Out	In the floating tidal city, the entire city rises and falls with respect to the sea bed. Pistons constantly at the town level press water in and out of hydraulic stanchion and the water operates the generators.	Concept.
	There are a vast number of geothermal pools where the temperature is insufficient to boil water but would boil an organic liquid. Were the resulting electricity to produce hydrogen, many of these pools could be connected by hydrogen-carrying pipes.	Concept only.

Part IV

Extrascientific Considerations

18

The Politics of Survival

INTRODUCTION

Politics? According to the dictionary, it means "practical wisdom," or "crafty and unscrupulous," or "expedient." What politics usually means in practice is the art and science of distributing the money which accrues to the government from the people's taxes. Politicians are the people who decide this distribution, who make the decisions as to where and how it is to be used, and thus, as the flow of money controls action, it is by and large the politicians (more than any other group in the community) who control what new projects, whether they be road building or research, are carried out.

This book has explored the vital and basic subject of the energy supply, and the impending energy disaster we are faced with, stressing the probable future of the next generation. *It is important to realize that the present generation will see the supply of oil and natural gas running out.* This book is about the things which we know to be relevant in making a new and different energy supply for the future, and we have shown that this essentially means getting abundant clean energy from sources such as the sun. The point we have been making is that these new energy sources must be *clean, abundant,* and *inexhaustible.*

To attain a new and more hopeful chance for the future energy supply, a radical turnabout in the direction of research and development for energy sources in the major countries of the world is essential. Available energy has been taken for granted in the past, coupled to the reserves of oil, natural gas, and coal. It was erroneously thought that these commodities would last too far into the future to require us to plan for tomorrow's energy. Research work in energy has gone in the past predominantly into atomic power, and more recently into the gasification of coal. Except for the study of fusion, little research money was applied in the direction of clean, inexhaustible energy sources before 1975.

The Direction of Major Research Funding Depends Upon Politicians

Research scientists get employed by industrial firms, government laboratories, and universities. The type of research they pursue is a matter of concern to all citizens. Is there a sufficient number of people working on the most pressing survival problem of the time: how to make the new machinery of collecting abundant clean energy ready *before* the fossil fuels exhaust?

Only a small number of scientists, engineers, and technicians employed by government and industry are actually directly involved in research and development work. Most of them do desk work, or continue to pursue and improve old, established ideas. A few do the innovative work which has potential for leading to a better future. Very, very few are allowed to just think.

The point that many people do not understand is that research and development for a new technology needs funding, and the use of tax money is essential for such purposes. The cost of doing fundamental research is two to four times the research worker's salary per person; the costs of the next stage, the development and pilot plant work, are much greater; and the costs of actually building the machinery which has been shown by this research and development to be practical are tremendous.

Given the limited amount of money spent on research, we must make certain that the right fields of research get government funding. Without this funding, nothing will happen in the way of any new technology.*

The decision of which research project of far-reaching significance is to be funded is made by committees of the federal government. The scientist sometimes has an advisory role to these committees, but he does not make the final decisions. *Scientists and government laboratories can only do what politicians allow them to do*, via funding.

---

*What about research done by industrial companies? Aren't they meant to produce new technology? They surely are. But the point here is that these firms are owned by people who have bought shares in them for *one reason:* to make some money. So, they are exceedingly interested in their company's scientists inventing, e.g., a tennis ball which can be used in the rain, or a color TV with better hues, etc. With such short term, consumer-oriented goals, they will see actual profit money in their pockets in 1–5 years after the invention. But what of their attitude to research on something great and wonderful, totally vital, at least ten, probably twenty, and possibly thirty years away in realization? Remember, investors are people. Pay off in twenty years? We may be dead then, and we want to live now.

What Does Survival Mean?

Talking about "survival" nowadays is considered too dramatic and in poor taste for a scientist. We are supposed to be rational, serious people, with an objectivity that should prevent us from becoming overly pessimistic about the future. Especially in industry, the researchers must keep a stiff upper lip and project the brightest version of the truth as they see it, in order to get promoted within their companies. However, concerning the future of the energy supply, it is difficult not to be dramatic! Therefore, as a conclusion to all the ideas and information presented in this book, we are going to present some questions, here and in the next section. Each individual citizen should not only be aware of these questions, but must also form positive attitudes about how to solve our energy exhaustion crisis, based on conclusions shaped by these questions. Switching the lights off when not in use and using less heating and air conditioning are good, but not enough. We must become actively involved in decisions our governments make about funding and about *our* future.

If we do not have machinery to tap abundant clean energy sources like solar energy before the fossil fuels run out, we shall necessarily have much less energy per head when they do run out. When energy supplies decrease, energy will get more expensive due to the short supply, and we will not be able to afford enough of it to supply us with what we think are our needs.

What would the resulting situation be? How many hours per day could we then afford to light our rooms? Could we afford to have convenient heating? At first it would be thermostats at 18°C (65°F), then 15°C (60°F), then back to wearing overcoats indoors at 10°C.* How many individuals could afford to heat their homes at, say, five times the present price of oil in dollars? Should we heat just one room well or the whole house poorly?† What would we be able to afford?

What about traveling in cars? Could we afford to travel in private cars when gasoline costs 10 dollars per gallon? How about public transportation? How frequently would that run? Would it run at all except for getting people to work and home again? Already, fleets of Boeing 747 planes are in mothballs for lack of use at the present cost of fuel. And how many days per week would the work places then be able to stay open—three days in the week? Two?

Now how about the distribution of food as the price of energy increas-

---

*Wood burning would last only a short time. Imagine how many fireplaces would need large quantities of wood per winter in one major city alone.

†Only 50 years ago, in Europe, before cheap oil became available, it was customary to have a single room per house, the "living room", heated in winter.

es? How would food be processed in factories when the energy which runs such places would have become too expensive to use? How about medical services and supplies? And at what degree of energy shortage (i.e., at what point in the rise in price of energy) would service systems begin to break down? What about trucks, trains, and high-energy using planes (forget *them*) which are used to bring supplies?

Enough said. Imagine the rest. A sufficient shortage of energy, which means "sufficiently expensive energy," would cause life as we know it to deteriorate. The economy would stop functioning, as would life, first in the biggest cities, and then everywhere else in the industrialized world.

Some people might say "Good, so we'll go back to the land." What land? Ten million people in New York City area would go back to the land? Where? Food? Drainage? Medicine? Training and supplies?

The Idea of Vested Capital

Let us back up now and take a look at the system under which we live, and which is threatened by energy exhaustion. We are talking about the free-enterprise capitalist society. The possession of capital (whether it is owned by a person or a company of private individuals, or the state) is investment. The owners of the capital buy something with it which they hope to turn into profit for themselves.

The Tobacconist

On a small scale we can see this in the example of a man who buys a tobacco shop. He pays some money (inherited, saved, or borrowed) for the tobacco shop, i.e., he invests in it. Having bought it, he hopes that he will be able to make some money out of it. Not only a living, but more than a living, so that, eventually, he can buy a second tobacco shop, a third, and so on.

In this ideal case, the business person is building up his capital, since each of the tobacco shops will produce more capital with which he will be able to invest and buy more tobacco shops. Eventually, ideally, the investor will be able to sit back, appoint a manager who will look after all the tobacco shops, putting a person in charge of each one, with assistants, and the little empire will multiply and grow.

Our buyer of one tobacco shop has now grown into a successful capitalist. He can work a bit less, and perhaps worry a bit more, but he can now afford to buy many pleasures which will soon seem to him to be necessities—cars, a small plane, several homes and apartments, etc. Now, all will go well for the man if he continues to work sufficiently hard and checks up on his managers diligently. Unless one dreadful possibility occurs—suppose the government puts a restriction on the sale of tobacco.

Now the capitalist who invested his money in the first tobacco shop, and made it grow, is in a dreadful state. He has to close down his shops. He cannot recoup his invested capital because nobody wants to buy tobacco shops anymore. They are no longer a viable proposition for money-making and have become valueless. He has suffered disaster. All his hard work, his 14-hr workdays, in building up his empire, has amounted to nothing.

From this simple example, an essential concept emerges:

Capital, vested in things, has an inner resistance to change.

This resistance is not dependent upon the machinations of selfish and sinister people behind the scenes, portrayed in cartoons as overly plump, older men, smoking cigars and smiling nastily. It is the resistance of ordinary people who have worked abnormally hard to get ahead. It is also the resistance of the people associated with them who do not wish to lose their jobs, for in capitalism, it is then up to them to find another job (although they were layed off through no fault of their own). The times without jobs are very rough, indeed, and not at all good for life at home, where the standard of life has to be suddenly reduced many times.*

So capital invested in industry has a degree of continuity all its own. And so does the type of industry and energy economy in which the capital has already been invested. No new ideas of scientists will be enthusiastically developed whilst the capital needed to do this is earning a good return on old ideas, especially when the loan from the bank with which the equipment for the older technology was bought, is not yet paid off.

THE POLITICIAN'S DILEMMA

The first law of politics is: "be re-elected."

In a democracy ("the rule of the people by elected representatives of the people"), people have an effect upon the government. In general, they do not re-elect a politician who they consider to have brought about measures which decrease their living standards.

---

*Reduction in living standards is never easy, as was evidenced in the world's Great Depression of the 1930s. For children born after World War II, at least in the U.S., it is taken for granted that the standard of living will always increase.

A wise politician, seeing the scale of time in the energy picture (see Chapter 4) would plan for the people's future needs. He is, after all, supposed to look after their interests. This means that he will tax the people *now*, to fund the necessary technological development for a safe future. Because of this taxation, his constituents who voted for him with the aim of getting bigger cars and better vacations, "the good life," will have to instead cut back their hopes and expectations. This is necessary so there can be the development needed, a life for tomorrow.

But people, by and large, do not realize the need for future investments when the payoff is decades away. They do not realize that research and development for a new energy technology take decades and need thousands of scientists and engineers, costing very large sums in tax dollars. They are more interested in today than in tomorrow. In our social system, the people do indeed influence the government, and they can prevent the politician from being re-elected if they do not like what he has done for their standard of living *now*, i.e., the degree of material satisfaction he has enabled them to have. But, if he is not re-elected, he will no longer be an active politician, and will lose all of his influence. Hence, he will often be savvy enough to work for the short term. Intelligent and idealistic at first, he will probably learn that the worst thing that can happen is not being re-elected, for then he will not influence things *at all*. So he aims to please people in his first term, hoping to be re-elected with the help of some funds from some of the groups he has helped. He must then please people again, and hopes someday to be able to do something about long-term goals.

Thus, one penalty of our system of government is that it hampers long-term planning. It is really only practical to please people from term to term, never looking beyond the next election. This provides planning for 2–4 years ahead, but very often the time scale necessary is decades, as is the case with our energy situation.

Economies Cannot Expand Forever—When Will Growth Stop?

Much has been published since 1972 about the necessity to limit the growth of the world's population and industrial production. However, anyone who applies some thought and common sense to the problem should recognize these limitations on their own. The eventual cessation of the growth of population is inevitable, because of the limited size of livable space on earth. If we build more and more houses, using more and more land, we will simply fill them with more and more people. If we find more livable areas, e.g., expand to other planets, we will assume it simply justifies further expansion, just postponing an inevitable dead end. Unless our attitude of continual growth stops, we will always expand to fill all available space and need to search once more for new inhabitable places.

So, we have to stop the growth of populations. This conclusion comes very much in conflict with the wishes of those who possess capital. If a person owns a group of apartments, she wants the surrounding area to grow so that her living accommodations will be in demand. Then, rents will rise, and the buildings will be worth more. Growth and expansion have become synonymous with prosperity and well-being. So, although people may recognize the necessity of reversing the economic expansion trend, they are more than willing to postpone it to a time when it will no longer affect them personally.

ENERGY DISASTER?—PEOPLE, POLITICS, GOVERNMENT FUNDS, AND RESEARCH

Big research money is allocated by government committees. But members of the committees are often employed by companies with a large investment in older technologies

We have stressed that funding is decided by politicians, who, in turn, are influenced by a desire for re-election, which will be determined by their ability to please the public, and those who control money which can be used to influence the public (e.g., big businesses).

If you sell shoes, you know about shoes, and you may not have the time to learn about energy—the pros and cons of what research has to be done to avoid an energy disaster. So you gladly leave it to your congressperson, who belongs to a committee which decides what research is to be supported with funds (your money from taxes). Hence, much depends on whether the congressperson knows a lot about the situation, and whether he or she will use this knowledge to *honestly* represent the long-term interest of the people. In general, the representatives will be more responsive to the short-term aims of corporations and the investors, who finance their election campaigns if pleased. And it is these money donors who have everything at stake, vested in the *present* technology, which we have shown to be resistant to change (as with the tobacconist). The companies with interests in energy resources want maximum profits and control for oil exploration and exploitation, and will never welcome an end to their monopoly of energy resources, where the key lesson is: shortages and fear are good for profit.

This brings up the sad subject of corruption. As in any profession where power is held over people and money, politics is sometimes the scene for decision-making on bases other than the good of the people. Instead, the determining factor is indeed the good of the person making the decision. For example, perhaps a company representative expresses to a certain congress-person that the company's board of directors is so delighted by the politician's efforts that they will put a company plane at his disposal so he can get around a little more, a little faster, and in a little more comfort. Much later, they may gently express the hope that the politician will vote their option on some upcoming legislation. If he does not, what will happen? Perhaps, nothing. But the politician knows he will not be rated so favorably in the company's eye when budget time rolls around, and he may not receive the indirect financial support previously proffered. Corruption can be gentlemanly.

Indeed, corruption might influence some politicians to fund work which would encourage the use of the old sources of fossil fuels, due to the monetary influence of the fossil-fuel owners who naturally wish to keep consumption and profits at a maximum whilst the fuel lasts, and can afford to coax a few politicians into favorable legislation. This could mean that research funding will not be given in time to develop the new energy sources, in which no one's money is yet invested.

We must observe, then, and understand, these extrascientific influenes on the futures of us all. They are part of the costs of our system.

19

Answers

What energy crisis?

This refers to the impending decline of living standards which will continue until renewble resource energy machinery is built.

Is oil really running out?

Yes. It is a misconception to think that, with a rise in price, more oil will always be available. The production rate of world oil will climax in the early 1990s. The subsequent drop would continue without end.

What is energy?

It is that which causes change. The realization of how energy manifests itself in movement and work is given in Chapter 1.

Can pollution be avoided?

Yes. Clean energy sources are a possibility; for example, they could be available to us if we took our energy from the sun, and to some extent, from winds and tides, using a fuel such as hydrogen. But, the cleanest may not also be the most economical way to provide energy; it is dependent on the cost. It depends on how much one is willing to pay to obtain an energy source which can be used for the foreseeable future and which will not have a finite lifetime, because it would make the world too dirty to live in.

Is fission safe?

It may be a more dangerous option than the use of coal, or, of course, the use of solar energy on a massive scale.

Are breeder reactors the answer?

This is the most dangerous option.

Fusion?

Maybe. But not soon enough to take over when other sources run out.

Is solar energy feasible and practical on a massive scale?

Yes. It may be somewhat more expensive than some other options. It will, however, give us clean and safe energy forevermore.

What about coal?

It would give us dirty air for some 50 years, and then decline.

What is a renewable energy source?

One which has effectively infinite capacity, for example, the sun.

How long will it take to develop alternatives?

Decades, perhaps 50 years for total conversion, for ideas which are already developed scientifically and need only technological work.

How much will it cost?

About two extra mortgages per family for 50 years.

How much time is left before the alternatives must be in full use?

This depends upon how much living-standard reduction we are willing to undergo. But, within two generations, we would have to be substantially converted to alternative energy resources.

What effect will the energy crisis have upon our present lives?

A reduction in living standards.

What effect will the impending energy disaster have on the lives of our children?

In the best case, we can recover, and a rise in our living standards can begin in a few decades. In the worst case, we will return toward an agricultural or primitive civilization: one that might support one-tenth the present population. The rest of the population would be destroyed by war, hunger, and disease.

Glossary

CHAPTER 1

Energy: This is a concept which is convenient to use in explaining many phenomena. One simple, short definition of energy would be: "That which causes motion."

Kinetic energy: This is the simplest kind of energy, the energy of motion.

Potential energy: This is stored energy, energy waiting to be released.

Photosynthesis: The formation of cellulose by the light-energized reaction of atmospheric carbon dioxide and water.

Hydroelectric power: When water falls and its kinetic energy turns turbines linked to electrical generators, the electricity is called hydroelectric power.

Transducers: These are machines which convert one form of energy to another. For example, an electrical generator uses mechanical energy to turn its armature inside a magnetic field; it produces electricity. It has transduced (or converted) mechanical to electrical energy.

CHAPTER 2

Feedback: A consequence of something done. For example, hard work often feeds back as high income. Environments at high temperatures feed back as slow and unenthusiastic work. High educational levels feed back as high earning capacity.

Exhaustion (of fuel supply): Simply, this means a situation where a given fuel, say oil, is used up so that no more can be extracted from the ground. In effect, we shall never reach this state, because long before we do, the price of oil will be so high that it will be cheaper to buy energy from another source. We shall say that oil has become "uneconomic" as an energy source. As the literal exhaustion approaches, there will be a considerable rise in price and this will continue (depressing living standards) until the machinery for the extraction of energy from another source has been built (whereupon the living standard should rise again).

CHAPTER 3

Combustion: We get energy by burning fuels like oil (a mixture of hydrocarbons, typically octane and decane). When we combust these oils, heat is given out and gases expand, exerting a force on that which is in their way, as with a piston in a cylinder.

Internal combustion: Combustion in a confined space as in the cylinder of a car. The expanding gases cause the piston to push, and thus a force moves through a distance and work is done, energy used.

External combustion: Combustion not in a confined space. The external heat makes a liquid vaporize in a confining "boiler" and the pressure thus exerted moves things, as the piston in a steam engine, which may drive a generator of electricity.

Fossil fuels: Coal, oil, natural gas. Called *fossil* fuels because they are old deposits removed from the earth. They were first formed by photosynthesis, i.e., the action of solar light on water and carbon dioxide.

Energy media: Substances (e.g., hydrogen) used to ferry energy around from source to user points.

CHAPTER 4

Resource: Something which is the origin of a necessary supply of material. Natural resources are the mineral deposits, including coal and oil. Human resources are, e.g., the number of scientists trained in energy fields.

Maximum production rate: As an economy expands, it will use up a resource faster and faster. Up to a point there is always enough of the resource left to be able to supply the entire amount, so the rate of production of the resource can respond to demand and increase. But, eventually, the demand becomes too much for the remaining resource and the possible production per year rapidly falls off, thus passing through a maximum production rate.

Living standard: The degree to which a community may have available to it, per average person, materials (e.g., housing, transportation), food, and services. The availability of nonmaterial things like leisure and lessened tension is beginning to be considered in the weighing of living standards (e.g., in the U.K.).

Strip-mining: In some parts of the world, e.g., the Western U.S., coal seams reach the surface and can be recovered by stripping off the surface cover, not by digging into the ground to find the seam ("mining").

Energy demand: As the aspirations and activities of a community increase, the energy it demands increases, too. This can either be met by a discouraging rise in price (which will slow down expansion) and/or encouraging increase in the means of production.

Growth rate: For many decades, all except the primitive communities have been *growing,* in respect to population, living standard, and energy demand. This can be described as a *growth rate,* e.g., 3% per year. The growth rate of Western economies in dollars which have been corrected for inflation ("real $") has been a varying amount, e.g., 2, 5, and 8% per year, for many years. However, energy has been plentiful, and it will be so no more until we have the machinery which can extract abundant clean energy from the inexhaustible sources. We don't know if we can sustain the growth rates of the past. Conceivably, the growth could become negative.

CHAPTER 5

Fundamental research stage: For some decades, there has been a stage where the fundamental physics, chemistry, and perhaps metallurgy of a problem is studied.

Development stage: Engineers build the early stages of a technology based on the fundamental scientific advances made by the scientist.

Commercialization stage: This is connected with converting something, which teams of engineers have made to work in trial plants, into something the public will want to buy.

Investment capital: This is the money, usually borrowed from banks ("mortgages") or the public ("bonds") which is used to build up the plant, etc., and exploit the invention.

CHAPTER 6

Fission: The breaking up of the atomic nucleus into other atoms and elementary particles.

Breeding: The use of neutons from the active (rare) form of uranium-235 to cause a nuclear reaction in several atoms of inactive (plentiful) uranium-238, for each neutron from uranium-235. The inactive (plentiful) uranium-238 nucleus becomes radioactive plutonium, so that much active (energy producing) substance is made from the small amount of (active) uranium-235 and the large amount of (inactive) uranium-238. This is *breeding* new atomic fuel (i.e., making radioactive atoms which give out energy from nonradioactive ones).

Forms of uranium: Uranium ores contain an oxide, and the atoms of this oxide contain two kinds of uranium whose weights are different. The 235 form is radioactive, undergoes fission and produces immense amounts of heat. The 238 form is stable, i.e., does not undergo fission. Uranium-235 has to be extracted from the ore for use in a fission reaction. Uraninum-238 is being saved for possible use in a breeder program if the breeder technology can be made to work, and if safety measures are approved.

Chapter 7

Fission reactor: An atomic device which produces heat when an atom of uranium breaks in two. The amount of heat energy produced per gram for this type of atomic energy producer is about 1 million times more than would be produced when a molecule of the compounds which make up oil reacts with oxygen in the atom.

Melt-down: If the cooling devices which keep the core of an atomic reactor from melting the surrounding container fail, the latter would melt and the core of the reactor fall to the bottom of the reaction chamber and, conceivably, pass further down into the earth. A melt-down would be extremely dangerous because it is likely that secondary (nonatomic) explosions would occur which might break the container and spew out radioactive atomic debris into the atmosphere. Were such a disaster to occur in a populous district, many thousands of lives would be lost in the first few hours.

LOCA: This phrase means "loss of cooling accident" and refers to the condition which would cause a melt-down as described above.

Dose rate: This refers to the time in which a given amount of radiation is absorbed by the body. It has been found that it is not only the amount of the dose which is important, but the rate at which it is received. Unexpectedly, the longer the period the dose is administered, the greater the damage per dose.

DNA: This is deoxyribonucleic acid. It is that molecule, which, within the body, contains the genetic code which decides how the various cells of the body grow. In normal life, cells reproduce themselves constantly in the same shape. In some conditions, radiation damages cells containing DNA. The result of this is that the code of these cells is changed; the cells grow in an undesired shape, i.e., become cancerous.

CHAPTER 8

Fusion: The joining by means of nuclear reactions of two nuclei to form one new (bigger) nucleus. Very much heat is given out.

Plasma: At sufficiently high temperatures, molecules break up into atoms and at still higher temperatures, the atoms lose some of their electrons. The system is then a mixture of ions and electrons and is called a plasma.

Magnetic field: At such high temperatures (millions of degrees Celsius), plasma cannot be held by material bodies which would promptly burn up on contact. It might, however, be held by a *magnetic field* in the shape of a container.

Controlled way (in respect to fusion): Fusion of hydrogen nuclei to form helium nuclei has been attained momentarily, and also explosively in the hydrogen bomb. The question is: how can it be maintained for long periods so that heat for electrical energy can be withdrawn from it?

CHAPTER 9

Galaxy: One of the ways we can regard the units of our existence is in terms of galaxies. Galaxies make up the universe. Our own sun is one of about 10^8 suns (or stars) in our own galaxy, the appearance of which to the naked eye on a moonless night is sometimes referred to as the "Milky Way."

Z.P.G.: Zero population growth. This is an ideal, already reached in Japan. The best situation would be Z.P.G. and an increase in energy per person. The worst situation would be population growth and no corresponding energy supply growth. If the energy supply fell while the population grew, living standards would fall, eventually causing the end to life in towns, because these depend on a high energy intensity distribution system.

Insolation: The receipt of energy from the sun.

CHAPTER 10

Radiation: When a body is sufficiently hot, it emits a significant amount of heat in the form of radiation.

Absorption: When light strikes a body, it will be either absorbed or reflected or (usually) both. Absorption means that the light energy is "taken in" or "accepted" by the substance on which it falls. Such a substance will get hot—it is the light energy which is providing the heat energy. Reflection of light can also occur—the photons bounce back and strike our eyes so we see the body. A body which absorbs all and reflects none looks black. A body which reflects all the light incident on it will look white.

Photovoltaic: A substance A, which when joined to a substance B—the couple being exposed to light of a suitable wavelength—sets up an electrical potential difference between A and B. If there is a wire connecting A and B, an electric current will flow, i.e., the energy of the light from the sun can be tapped.

Ocean thermal energy collector: A method for converting heat to electricity. The upper surface of a tropical sea boils a low-boiling liquid and the vapor from this works a heat engine. Cold water is pumped up from the depths and condenses the vapor in a condenser to a liquid. This is evaporated in a boiler by the warm water again, and so on.

Power tower: Concave mirrors surround a tower. Each focuses the sunlight which it receives onto the tower top, where sits a boiler. The boiler gets hot and the expanding vapor from the liquid in it drives a heat engine and can then drive an electricity generator.

Cadmium sulfide: This is one-half of a couple consisting of cadmium sulfide and cuprous sulfide. But the term is used to mean the photovoltaic couple consisting of the two sides of the couple.

Silicon: In the photovoltaic sense, a silicon photovoltaic couple consists of two parts, one of *p*-type and the other of *n*-type silicon (see text). Joined and then irradiated under direct strong sunlight, they give a 0.8-V electrical potential difference. An electric current can be drawn from this system (as with other photovoltaics) and solar energy has been converted to electricity.

CHAPTER 11

Roof-top water heater: Black panels on a roof absorb the sun's radiation and become hot. When water is passed over the hot panels, it obtains some of this heat.

Black panel: This may consist of a metal oxide of dark color. Dark organic substances are usually decomposed by heat.

Ducted heat: The hot air from the roof-top solar heated panel is pumped through metal pipes ("ducted") to the house.

Eutectic storer: Mixtures of substances of certain ratios have minimal melting points and are used as substances to be heated up with hot air from the rooftop panels and cooled down to freezing, whereupon they emit latent heat. Air circulated over them can give the stored heat back to an area to which the air is then ducted.

Hydride storer: Heat can be stored in the eutectic storer. However, hydrogen can be stored, e.g., in a metal alloy, like iron–titanium. When the hydrogen is brought into contact with the iron–titanium at a certain pressure, the alloy absorbs hydrogen, and the hydrogen fuel is thus stored. When the hydrogen pressure is diminished, the alloy gives up the hydrogen it has previously absorbed.

CHAPTER 12

Electric battery: A device which uses electrochemical reactions at surfaces to *store* electricity put into the battery. On pressing a switch, the electricity can be produced again at will.

Fuel cell: A device which produces electricity inside itself when certain fuels (e.g., hydrogen and oxygen) are fed into it.

Energy source: Batteries are only energy *storers*. Cars run on batteries must have an energy source: it could be coal. Soon, it must be atomic or solar energy.

Recharge station: Cars working on batteries will *exchange* them at recharge stations for freshly charged ones. The exchange could be done in the same time as one now fills a tank containing gasoline. Recharging the rented batteries would occur at the recharge station.

Liquid hydrogen: One of the ways to run much of our present technology is to use hydrogen to replace natural gas and oil. The hydrogen could be stored in alloys such as iron–titanium, or contained as a gas in cylinders, or liquefied and stored in low-temperature tanks.

CHAPTER 13

Mean annual wind: Winds do blow irregularly, as is the common experience, but, if the mean of the winds is taken over the whole year, then — for a certain place—the wind velocity is constant. This means that we can rely on being able to obtain a calculable amount of energy from the wind at that place.

Aerogenerator: This is a device—usually propellerlike in form—which gives electricity when it rotates (working an electricity generator) under the force of the wind.

Rotor: The part of a wind generator which rotates. It may have several shapes, including a wheel-like shape (see Fig. 13.2) or a flag shape (see Fig. 13.6), apart from the propeller.

Storage: As wind is only of constant velocity in a given place on a yearly basis, and the times of need are quite different from the time at which the wind blows, energy has to be stored. Electrolysis of water and hydrogen storage is one way in which this could be done, whereupon the hydrogen could run cars. Battery storage is the usual way.

Geothermal energy: Energy obtained from the heat within the earth.

Earth's mantle: The outer rocklike cover which forms the external surface of the earth and contains within it a molten silicate.

Magma: The molten interior, 99% of the planet.

Low-grade and high-grade geothermal energy: High grade refers, in a geothermal energy context, to the heat within the earth's magma. At numerous points on the earth (distinct from volcanoes when *breaks* in the earth's mantle occur), the mantle gets thin. Nearby, the rock near the earth's surface gets hot. If an underground spring happens to exist in the neighborhood of the hot area, the water will become hot, sometimes hot enough to form steam and therefore become an energy source. This is called a low-grade energy source because its potential is not great (though, in New Zealand, about 6% of the electricity comes from this source).

Chapter 14

Raw energy: The energy resulting directly from solar or atomic (or gravitational, or geothermal, or tidal, or wind) sources; in contrast to the *derived* energy in the form of some fuel which we shall send around from the energy production plants and actually use.

Energy media: Forms of energy from raw energy and sent around to user sites.

Energy sources: These refer to the *origins* of our energy supply. Just as now our energy comes from stored solar energy in fossil fuels, so, in the future, our energy will come from solar and atomic sources.

Energy carrier: More or less the same as energy media, but stressing the transportation of energy, e.g., liquid hydrogen in a tanker is literally an energy carrier.

Ecology: Before man began to disturb the natural world which had been developed over billions of years, it ran in balance. Man has been disturbing this balance by developing technology, e.g., by having a transportation system which uses the atmosphere as the receptacle for the wastes (carbon dioxide, unsaturated hydrocarbons, etc.) from the internal combustion engine. As these wastes do not disappear but build up in the air, the balance of nature is disturbed. Continuation of such unbalancing could make air sufficiently modified to affect life.

Chapter 15

Cyclical method for getting hydrogen: In this method, one starts with a metal and ends with hydrogen and oxygen, regenerating the metal for use once more in decomposing hydrogen.

Electrochemical method for getting hydrogen: This involves electrolysis of water to produce hydrogen and oxygen, as described in previous chapters.

Chapter 16

Aquifer: This is a naturally formed cavity under the earth which contains water. This water is often hot (near to boiling) and is often unfortunately salty.

Flywheel: A wheel, generally heavy and fast rotating, which is often used to steady the motion of an engine in which pulses to energize it come at intervals. The flywheel's inertia of movement keeps the engine running smoothly. In an energy context, the meaning may be that of a wheel which, by means of a sporadic energy source, is made to turn at a high rate. The design will reduce friction to a minimum. The kinetic energy of the wheel is in fact a kind of energy storage.

Index

Abundant clean energy, its creation, 63
Abundant-solar-energy nations, why don't
    they act?, 123
Actuality of solar energy collection, 137
Aerogenerator, defined, 249
Aircraft, redesigned for hydrogen, 162
Alaska, and crude oil pipeline, 33, 34
Allis-Chalmers, and hydrogen produc-
    tion, 210
Alternatives, time left to develop, 242
Ammonia vapor, and its use in OTEC gen-
    erators, 136, 137
Answers to questions about energy, 241
Antarctic, and atomic wastes, 89
Aquifer, defined, 251
Atomic energy
  cheap?, 69
  and chemical energy, dreams fading, 71
  exhausting, too?, 75
  as a source, 65
Atomic reactors, the conditions to become
    explosive, 74
  and the Antarctic, 89
  and glassylike materials, 89
  and space, 89
  and tectonic plates, 89
Australia, and solar energy collection, 123
Australian desert, photograph, 123
Australian government, lack of research
    by, 142
Automobile, how it works, 27
Automobiles, electric, diagram, 154

Background radiation, and atomic power,
    85
Bacon, Tom, and the first practical fuel
    cell, 162

Barstow, California, and the 1982 plant,
    139
Batteries
  for cars, 156
  energy of charging, 158
  high-energy, 156
  sodium–sulfur, 157
Bay of Fundy, and tidal energy, 183
Big winds, where they exist, 171
Biological damage, and dose rate, 87
Biological hazards, mechanism, 86
Bipolar cell, for hydrogen production,
    210
Black holes, and futuristic ideas in energy
    production, 224
Boer, Professor K., and Solar I, 148
Bohr, and a British submarine, 69
Bottles
  containing plasma, diagram, 98
  and the escaping plasma, 96
Braking, regenerative, 19
Breeder reactors
  the answer?, 242
  diagram, 77
  drawbacks, 79
  and saving the situation, 77
Breeding
  defined, 245
  idealized, 77
  process, diagram, 78
Brody and Shirland, their estimates for
    photovoltaic energy, 131
Broome, Australia, and tidal generations,
    184
Brown's Ferry, 83
  and human error, 83
Building a new system, cost, 60

Cadmium sulfide, a photovoltaic material, photograph, 131
Cambridge University, and the first practical fuel cell, 162
Cancer
  and radioactive substances, 85
  time it takes to develop the disease, 88
Cancer-causing damage, 85
Cancerous glands, and distribution of iodine, 86
Capitalists, and ways of suffering disasters, 237
Carbon dioxide
  and the atmosphere, 165
  and climatic change, 166
  emitted from cars, 22
  the universal product, 35
Cargo aircraft, using hydrogen as fuel, 161
Carnot, his efficiency factor, 164
Carriers of energy, 194
Cars
  battery-operated, 156
  as users of energy, 18
Cavities, filled with hydrogen, 221
Cell, photovoltaic, 127
  for producing hydrogen, 210
Cell characteristics, in electrolyzers, 210
Cell construction, for hydrogen generation, 209
Cell damage, by radioactive particles, diagram, 87
Chemical energy
  conversion to kinetic, photograph, 8
  converted to electricity, 28
Chile, a future Saudi Arabia of wind energy?, 174
China syndrome, 82
Chlorine gas, and removal after electrolysis of seawater, 177
Civilization, and the desirability of the development of an interdependent energy network, 203
Coal?, 242
  exhaustion of, 49, 50
  illusions, 48
  less pollution if we produce hydrogen, 206
  its nature, 29
  the cheapest source of hydrogen, 205
  storage, 30
  strata diagram, 30
  use, 64

Coal? (cont.)
  will it last 1500 years?, 48
  will it last 50 years?, 48
Coal industry, its gigantic potential development, 64
Collection of energy, futuristic schemes, 227
Combustion engineering and mechanism, 26
Commercialization, the fourth stage in developing a future, 57
Compromise, importance of, 12
Concepts
  for the next few hundred years, 223
  for the next few thousand years, 223
Conditions to make space collection of energy economic, 134
Conservation, not the answer, 52
Control of people, 121
Conversion of energy to useful form, 10
Converter
  defined, 10
  importance of having an efficient one, 10
Cooling
  of hot rocks, 188
  by solar heat, 147
Corrosion, and atomic wastes, 89
Cost
  of building a new system, 60
  of developing alternatives, 242
  difficulty of estimating, in fusion, 99
  predominance of, as factor, 12
  of wind generator, tabulated, 180
Countries, and their energy per capita, plotted, 16
Couple, photovoltaic, 127
Critical experiment, of Petkau, 85
Curie, Marie, photograph, 70
Cycles, use in producing hydrogen, 206
Cyclical method, 207
Cyclicity, degree of, 208

Dakota, Alabama, and the LOCA, 83
Damage due to nuclear radiation, and genetic code, 86
Dams, generating capacity, 29
Dangers of using hydrogen, 221
Darkness, and the supply of solar energy, 149
Deactivation in photovoltaic crystals, 130
Deaths from atomic power, 85
Decades, time to think in terms of, 55
Decline of energy supply from fossil fuels, 22

Delaware River Valley, and transport of hydrogen to, 120

Democracy, and limitations of ability to act, 61

Detector, and distribution of iodine in cancerous gland, 86

Deuterium
its diluteness, 92
and role in fusion process, 105

Development times of various energy sources in the past, 60

Developmental engineering, the third stage in developing a future, 57

Dilemma for politicians, 237

Diluteness
of deuterium, 92
solar energy's Achilles' heel, 114

Dirt, and energy sources, 21

Disaster, energy, 1

Dismantling of civilization, possible?, 121

Distance, and solar energy, 119

Distances, and transmission of energy, 196

DNA
defined, 246
and future generations, 87
and involvement in genetic code, 86

Dome, would it hold?, 82

Dose rate
and biological reactions, 87
defined, 246

Dreaming
of atomic energy, 69
about fusion, 91
and Moscow, 56

Dumping, radioactive garbage, 89

Dyson sphere, 225

Earth
and its crust, diagram, 182
and moon, 182
thickness of its skin, diagram, 186

Ecology, defined, 250

Economies
end of growth, 238
and growth, 51

Economists, their unfounded optimism, 50

Economy, hydrogen, 201

Efficiency
of collection of solar energy, 115
of energy conversion, 11
improved, of electrochemical engines, 164
of solar collectors, 115

Einstein, and his warning to the American government, 69

Electric cars, 156

Electricity
and difficulty of transmission over long distance, 193
generating facility, photograph, 36
generating plant, oil-fueled, 35
generation from fusion, 103
generator, how it works, 28
for houses, and photovoltaic couples, 149
how it is made, 26
production from fusion, diagram, 103

Electric motors, overhauled only once per million miles, 203

Electrochemical energy producer, diagram, 165

Electrolysis, characteristics, 209

Electrolyzers, diagram, 210

Elliot and Turner, their work on fossil fuel exhaustion, 44

Emergency system, and lack of function, 84

Employment, and fossil fuel exhaustion, 45

Energy
after collision in chemical reaction, 5
carriers, 193, 194, 250
to charge batteries, 158
classification, 3
and comfort factor, 112
consumed by average person, 112
content of, in fossil fuels, 40
converted to useful form, 10
divided into various uses, 15
in the fossil fuels, 9
and future needs, 113
geothermal, 184
and industry, 34
and our lives, 13
needed by the year 2000, 113
network of, international, 204
per person, measured in kilowatts, explained, 113
percent in various products, 14
price of, 55
related to comfort and efficiency, 15
running out of, 39
sources, present ones, defined, 25
from sun, 107
supplied by breeders?, 79
and tides, 169
transmission over long distances, 196

Energy (*cont.*)
  and its transportation, 37
  types of, 16
  what is it?, 3
  where we get it from, 25
  and winds, 169
  from winds, and wind velocity, 171
Energy collection, future schemes, 226, 227
Energy content, of various things, 14
Energy crisis?, 241
  and our present lives, 242
Energy disaster, 239
  and the lives of our children, 242
Energy future, problems of, 36
Energy sources, finite, 38
Energy storage, 190
  in metal hydrides, diagram, 220
  methods, 216
  methods to be used in our times, 219
Energy use, commercial and industrial,
      examples, 19
Energy used in residence, subdivided, 18
Engineering, planetary, 224
Engines, ordinary, their lack of efficiency,
      163
Error, human, and Brown's Ferry, 83
Eutectic storer, defined, 248
Evaporation of seawater and solar energy,
      66
Exhausting fossil fuels, their replacement, 63
Exhaustion
  of atomic fuels, 75
  effect on living standards, 45
  and rise of living standards, 45
  what it means, 22
  when it will occur, 22
Exhaustion of fuels, dramatic effects, 235
Exhaustion rate, becoming exponential, 43

Fantasies, for energy production, 224
Farmer, French, 40
Fast breeder reactor, diagram, 76
Feedback, defined, 243
Fermi, and atomic fission, 70
Fields, of natural gas, 33
First technological race, its future, 55
Fission
  compared with fusion, 93
  and safety, 241
Fission chain reaction, diagram, 74
Fission reactors
  close calls, 81

Fission reactors (*cont.*)
  defects, 81
  explained, 72
  how they work, 72
  and the LOCA, explained, 81
Floating cities, and energy generation,
      229
Food
  from hydrogen, 165
  synthetic, and British Petroleum Com-
      pany, 167
Fossil fuels
  defined, 29
  exhaustion as a function of time, 44
  how much is left?, 39
  and how they got their energy, 9
France, an early developer of solar energy,
      138
Freedom and growth, 52
French farmer, 40
Fuel cells
  advantages, 213
  defined, 249
  and efficient conversion, 19, 163
  more efficient than other conversion me-
      thods, 213
  and hydrogen, 161
  and production of electricity, 213
  schematic, 163
Fuel oil, its rise in price, 151
Fuel properties, compared, 155
Fuels
  compared with hydrogen, 155
  potential energy in, 7
  used now, limited, 39
  useful, and solar energy, 125
Fundamental research, the second stage in
      developing a future, 57
Funding, dependence upon political consid-
      erations, 233
Fusion, 242
  actual attainment?, 94
  advantages, 92
  the best source?, 104
  compared with fission, 93, 94
  compared with other sources, 105
  difficulties of realizing, 95
  a dream displaced, 91
  and electricity generation, diagram, 103
  how long?, 104
  and initiation by intense compression, 101
  and the laser method, 101

Fusion (*cont.*)
  and magnetic bottles, 97
  and money spent, 105
  not to be developed for 50 years?, 141
  the possible alternative, 91
  role of deuterium, 105
  utopian?, 92
Fusion power plants, and laser-induced re-
    actions, 102
Fusion reactions
  diagram, 95
  and remaining tasks, 106
Fusion reactor, and 50 years, 106
Future, ideas for, 223
Future needs, of energy, 113
Futuristic schemes, for energy collection,
    227

Galaxy, defined, 247
Garbage, radioactive, where to dump it, 89
Gaseous storage, of energy, underground,
    219
General Atomics, and its work on fusion,
    100
General Electric, trying to confine a plasma,
    99
Geothermal electricity, and Wilson, 187
Geothermal energy, low-grade, 190
Geothermal energy, prospect, 190
Geothermal heat, and energy production
    from, 169
Geothermal pools, and energy generation
    from, 229
Geothermal reservoirs, wet, diagram, 189
Glass bricks, and atomic wastes, 89
Glossary of terms, 243
Going back to the land, not practical, 235
Government dictates, their limitations in a
    democracy, 61
Grandpa's Knob, 176
Gravitational energy, defined, 4
Great Britain, no massive collection of solar
    energy, 115
Great Southern Wind, 174
Gregory, his calculation of hydrogen trans-
    mission costs, 197
Grids, carrying energy, 37
Gross National Product, as function of en-
    ergy, plotted, 15
Growth
  of economies, 51
  the halting of, 51

Growth rate, defined, 245
Guidelines, on time to develop technology,
    59
Gulf Stream, and production of energy, 228

Hahn, and atomic energy, 69
Hazards, biological, of nuclear reaction, 84
Heat
  and electrochemical storage, compared,
    218
  geothermal, 67
  production of hydrogen from, 207
Heavy particles, and how they affect the
    body, diagram, 87
Helium, and fusion reaction, 100
Hildebrandt
  photograph, 119
  and the solar thermal method, 118
Hiroshima, and atomic energy, 69
Home heating, by solar energy, diagram,
    144
Homes, and electricity we use in them, 212
Hot rock geothermal energy
  diagram, 188
  its difficulties, 187
  practical?, 187
Hot water, production from solar light, 143
Household electricity
  from solar light, 148
  from winds?, 170
Household energy, 17
  from the sun, 141
Houses run by solar energy, and availability
    of solar energy, 151
Hydride storer, defined, 249
Hydroelectric energy, a solar-gravitational
    source, 66
Hydroelectric plant, 30
Hydrogen
  and cavities, 221
  cheap for long-distance transmission,
    196
  compared with other fuels, 155
  and costs of transmitting energy over
    long distances, 197
  cyclical method for producing, 205, 251
  and dangers of, 221
  drop of, and light beam, 101
  the ecological source, 203
  the electrochemical method, 208
  and wind energy, 169
  and food, 165

Hydrogen (*cont.*)
  fuel for spacecraft, 166
  gaseous, and getting energy back, 210
  and a hydrogen economy, 202
  and metallurgy, 165
  methods for getting, 251
  methods of mass producing, 204
  and planes, 160
  produced from solar energy, 118
  replacing natural gas, 201
  replacing oil in internal combustion
    engines, 165
  and transportation, 153
  used at Kennedy Space Center, 165
  used to run industry, 164
Hydrogen economy, 201
  advantages, 203
  ideas beyond, 223
Hydrogen energy storage, in metal
    hydrides, 219

Illusions concerning the lastingness of coal,
    48
Income, and energy, 15
Industrial research, lacking long-term
    goals, 234
Industrial world
  and the need for solar energy, 121
  receipt of solar energy in the form of
    hydrogen, 120
Industry
  how it gets its energy, 34
  running it on hydrogen, 164
Inertial confinement, and fusion, 99
Inflation and expenses, 49
Inheritance, of money, and fossil fuels, 23
Internal cumbusion engine
  and energy use, 26
  how it works, 27
  run on hydrogen, 165
Investment capital, defined, 245

Jet propulsion, and liquid hydrogen, 160
Justi, Professor E., a pioneer in hydrogen
    as a fuel, 201

Kennedy Space Center
  photograph, 166
  and storage of hydrogen energy, 217
Kilowatts, per person, diagram, 113
Kinetic energy
  description, 4

Kinetic energy (*cont.*)
  examples of practical values, 7
  obtained from potential energy, dia-
    grams, 8
  understood, 6
Kordesch, his fuel-cell–battery hybrid, 155
Kummer and Weber, their sodium–sulfur
    battery, 159

Lasers and fusion, 101
Lastingness of coal, explained, 48
Lead–acid batteries, why use?, 156
Life style, and living on capital, 23
Light, and origin of energy in bonds, 9
Liquid hydrogen
  and energy for planes, 160
  and storage of energy, 217
Liquid hydrogen storage, 217
Liquids, considered for OTEC plants, 136
Living standard, 14
  defined, 244
  and energy, 13
  and exhaustion of fossil fuels, 45
  and growth, 51
  reduction, 242
  and time, 46
LOCA
  defined, 246
  explained, 81
  and what could happen, 83
Looking ahead, how far?, 65
Low-mass particles, and how they affect
    the body, 87

Magma
  defined, 250
  in earth, diagram, 186
Magnetic bottle, and fusion, 97
Mass production of hydrogen, 205
Material damage, and neutrons, 97
Materials research, and damage from
    fusion generators, 98
Maximum production rate, defined,
    244
Maximum of resource production, 46
Media, 194
  pros and cons, 195
  and sources, 194
Meitner, Louise, and atomic energy, 69
Melt-down, defined, 246
Metal hydrides, and energy storage, 219
Metallurgy, and hydrogen, 167

Methanol, cost for production, 195
Microbes, oil-eating, 20
Mines, opening new ones every day necessary?, 64
Mirror concentrator method for solar collection, 134
Model for ocean thermal energy collection, 136
Moomba, and natural gas in Australia, 32
Moon
and earth, 182
and tides, 182
Moscow, statue in, "First one must learn to dream," 56
Mullett, Les
his design of wind generators, 172
photograph, 180

Natural gas
and electricity, 213
running out, 49
Neutrons, and material damage, 97
New energy source, and 50 years to develop, 60
1972, the year of change for photovoltaic cells, 130
Normandy, site of first practical tidal generator, 184
Northern Hemisphere, and land mass, 173
Nuclear reactor, photograph, 73, 82

Ocean thermal energy collector (OTEC)
defined, 248
explained, 135
Oil
crude, U.S. cycle, 41
and how it underlies our growth, 53
and living standards, 53
in pipes, 37
and price rise, social effects, 151
running out of, 49, 53, 241
where it is found, 30
Oil storage depot, at Russ-El-Tunurra, 215
Orbiting collectors, 132
photograph, 132
OTEC (ocean thermal energy collector), 135
defined, 248
diagram, 137

OTEC (cont.)
reverse collection, from cold mountain air, 227
why the delay?, 137

Pearson, Chapman, and Fuller, inventors of the solar cell, photograph, 128
People
and degrees of control needed, 121
and energy disaster, 239
lack of realization of need to plan ahead, 238
reaction to government dictates, 61
Petkau, an important physicist, 85
Photosynthesis, as greatest solar energy store, 215
Photovoltaic cells, and solar energy production, 118
Photovoltaic collectors, in orbit, 132
Photovoltaic couples, and energy for houses, 149
Photovoltaic generator, explained, 128
Photovoltaic method, explained, 127
Photovoltaics
characteristics, 129
thin film, 131
Pipeline from Alaska, 33
Planes, and hydrogen, 160
Plasma
in bottles, trying to escape, 98
defined, 246
described, 95
strange behavior, diagram, 98
Political considerations, in funding research, 233
Politicians
control of research, 233
and dilemma, 237
wisdom, 238
Politics
relevance for scientists, 233
of survival, 233
Pollutants, 189
Pollution
and our atmosphere, 20
avoidance of, 241
in cities, diagram, 21
and the high-energy life, 20
increasing?, 51
metallurgy, and hydrogen, 167
need to avoid, 63
and sulfur dioxide, 64

Potential energy
  in chemical bonds, 6
  conversion to useful energy, 6
  description, 4
  of various masses above earth, 5
Power relay satellite concept, diagram, 199
Power tower
  schematic of, 134
  what it would look like, 118
Practicality of wind energy, 177
Price, and its change, feedback effects, 47
Problems of the energy future, 36
Production
  rate, maximum—the effective end, de-
      fined, 42
  and its variation with time, 45
Productivity, its increase with time, 43
Proteins, production from hydrogen, 166

Questions, 1

Radiant energy, collection by panel, 127
Radiation, damage from, 85
Radioactive substances, causing cancer, 85
Rance, tidal generator at, 184
Realization of exhaustion, 9
Reflectors, for power tower, 135
Research
  and energy advance, 239
  and industrial companies, 234
Resources
  production as a function of price, 47
  sudden exhaustion, 42
Reverse OTEC collection, 227
Rotor, in Rance, diagram, 184
Russ-El-Tunurra, 37

Salt dome, and disposal of radioactive
      wastes, 87
Salt mines, and atomic wastes, 89
Satellite
  collector, in outer space, and energy col-
      lection, 227
  and pictures of air pollution, 21
  transmission, a space-age concept, 200
Schemes for future energy collection, 226
Scientists and politics, 233
Sea, and stronger winds, 174
Sea-born wind generator, 174
  diagram, 178

Seaborg, Glenn, discoverer of plutonium,
      77
Silicate, molten
  as heat source?, 186
  and interior of earth, 186
Silicon photovoltaics, characteristics, 129
Small rotor alternative for sea-born gener-
      ator, 179
Snowy River scheme, in Australia, dia-
      gram, 66
Sodium–sulfur batteries, 156, 158
Sodium–sulfur cell, 156
Solar alternative, too late?, 121
Solar collection, 115
  from orbiting platform, schematic, 133
Solar collectors, time needed to make, 120
Solar cooling, schematic, 147
Solar energy
  after dark, 149
  actualization of, by NASA, 138
  available before the oil runs out?, 151
  calculated for the whole planet, 116
  conversion to useful fuel, 125
  at far-distant points, 119
  feasible and practical on massive scale?,
      242
  future supply, 117
  how to get it during the night, 149
  lack of collection in Australia!, 123
  lack of development in the past, 123
  and the medium of its use, 117
  per person, 111
  and photovoltaic cells, 118
  reaching earth, 110
  for residential buildings, diagram, 150
  solar heating, 144
  in space, 137
  and space heating, 143
  and total energy needs, 115
Solar heater, diagram, 145
Solar household energy, 141
Solar houses, at the University of York, 149
Solar-hydrogen economy, diagram, 130
Solar lands, lack of action to collect solar
      energy, 123
Solar light, and space heating, 146
Solar sources
  direct, 65
  indirect, 65
Solar spectrum, 125
Solar thermal method, 134

Solar thermal method (*cont.*)
   what it would look like, 118
Sources of energy, 25
Sources and media, 194
Space
   and dreaming, 56
   orbiting laboratory, supplied with solar
      cells, 137
Space cooling, by solar light, 146
Space heating of houses, by solar energy,
   143
Space stations, collecting solar energy,
      schematic, 133
Spacecraft, running on hydrogen, 165
Spectrum
   distribution, 126
   solar, 125
Stack module, for hydrogen production,
   212
Standard of life, and energy, 12
Stars, and plasmas, 108
Storage
   of abundant clean energy, 215
   of conventional sources of energy,
      31
   of energy, 190
      in the form of heat, 217
      methods, 216
      thermal and electrochemical methods
         compared, 218
   of solar energy, 215
   of wind energy, 174
Storing energy, and electrical form, 218
Strip mining, diagram, 33
Suddenness, the characteristic of resource
      exhaustion, 42
Sun
   described, 109
   diagram, 108
   expected lifespan, 107, 109
   fraction of radiation intercepted by
      earth, 109
   the most available source, 107
   photograph, 110
   swathe of radiation intercepted by earth,
      111
   total energy generated, 109
Surface of earth, 114
Survival, meaning, 235
Swiss government, their unanswered ques-
      tion, 82

Table, comparing fusion and fission, 94
Tasks which remain, in attaining fusion re-
      action, 106
Technology
   costs of various stages of development,
      58
   developing it, 56
   new, and time to build, 55
   for solar energy development, and past
      doubts, 123
Tectonic plates, and atomic wastes, 89
Telkes, Maria, photograph, 142
Thermal chimneys, and deserts, 228
Thermal methods of producing hydrogen,
      disadvantages, 207
Thermoelectric generators, and hot rocks,
      227
Third technological race, and blowing up
      Venus, 224
Three Mile Island, how many people will
      have been fatally affected?, 84
Tide
   and energy production from, 169
   and moon, 182
   10-m high, 182
Time
   of development of alternatives, 242
   and the development of fusion, 104
   to develop technology, guidelines, 59
   left, vital nature of, 12
   needed, to make solar collectors, 120
Tobacconist, 236
Transducer, defined, 10
Transmission, of energy, 193, 197
   over *very* long distances, 198
Transport, run on electricity, 153
Transportation
   of energy, 17, 37
   private, and "freedom," 153
Tritium
   and role in fusion process, 105
   struck by laser, 101
Turbo fan, diagram, 11

Underground storage of energy, 219
Unemployment, and fossil fuel exhaustion,
      45
United States, and origin of oil supply, 50
Uranium
   sources, 75
   two forms, 73

Uses of energy, 15
Utopia, with fusion, 92

Vehicle, and where to put the fuel, 159
Venus, blown up, 224
Vested capital, 236
Volcanoes, as energy sources?, 185

Wairacki, New Zealand, and the equipment
    for geothermal generation, 190
Water
  an energy storer, 149
  as a heat transfer agent, 149
Western oil, in our children's lifetime, 54
Wheeler Dam, diagram, 29
Wilson, Dr. S. H.
  and geothermal electricity, 187
  photograph, 186
Wind belts
  and energy at sea, 177
  importance of discovering, 173
Wind energy
  conditions for success, 170
  estimates, 171
  and fossil fuels, 175
  stored, 174

Wind farms, on the sea, 228
Wind generators, 172
  in Antarctica, 228
  cheapest source of energy?, 181
  and the necessary radius, 172
  sea-born, 175
  at sea, diagram, 180
Wind map, of United Kingdom, 173
Wind power, massive, a practical propo-
    sition?, 177
Winds
  big, and energy production from, 169
  and household electricity?, 170
  made reliable by the use of hydrogen,
    169
  at sea, 174
Wood burning, not so good, 235
World population, estimate, 23

Years of doubt, 122
Years left, for main sources of energy, 53

Zero growth, its achievement, 62
Z.P.G. (zero population growth), de-
    fined, 247